# ENERGY EFFECTIVE INDUSTRIAL ILLUMINATING SYSTEMS

## DESIGN AND ENGINEERING CONSIDERATIONS

# ENERGY EFFECTIVE INDUSTRIAL ILLUMINATING SYSTEMS

## *DESIGN AND ENGINEERING CONSIDERATIONS*

### KAO CHEN, P.E., FIEEE

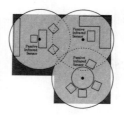

Published by
THE FAIRMONT PRESS, INC.
700 Indian Trail
Lilburn, GA 30247

**Library of Congress Cataloging-in-Publication Data**

Chen, Kao, 1919-
    Energy effective industrial illuminating systems / by Kao Chen.
    p. cm.
    Includes bibliographical references and index.
    ISBN 0-88173-168-4
    1. Electric lighting.  2.  Industrial buildings--Lighting--Energy conservation.
3.  Electric power--Conservation.  I.  Title.

TK4399.F2C49        1994        621.32'254--dc20        94-14242
                                                                                    CIP

Published by The Fairmont Press, Inc.
700 Indian Trail
Lilburn, GA  30247

Printed in the United States of America

10  9  8  7  6  5  4  3  2  1

ISBN 0-88173-168-4    FP

ISBN 0-13-147380-8    PH

Distributed by PTR Prentice Hall
Prentice-Hall, Inc.
A Paramount Communications Company
Englewood Cliffs, NJ  07632

Prentice-Hall International (UK) Limited, London
Prentice-Hall of Australia Pty. Limited, Sydney
Prentice-Hall Canada Inc., Toronto
Prentice-Hall Hispanoamericana, S.A., Mexico
Prentice-Hall of India Private Limited, New Delhi
Prentice-Hall of Japan, Inc., Tokyo
Simon & Schuster Asia Pte. Ltd., Singapore
Editora Prentice-Hall do Brasil, Ltda., Rio de Janeiro

To May Yu, Jennifer,
Arthur, and Carlson

*Thanks for all the support
and encouragement.*

# Table of Contents

# Foreword

Since the energy crunch years, lighting retrofit has been one of the most popular projects for energy conservation. More often than not, the reported successful retrofits are based on the energy/cost savings or the number of payback years as the sole criterion. These reports rarely mention improvements in workers' visual performance. Consideration of this important and achievable goal should be the prime objective of any lighting retrofit project. This objective can often be achieved along with significant energy/cost savings if illuminating systems design principles are properly observed and applied to the project in its initial stages.

Lighting retrofit projects provide illuminating engineers a great opportunity to improve the lighting quality and enhance workers' visual performance and productivity in spaces where inefficient, poor quality lighting systems now exist. About three years ago, the EPA announced the Green Lights Program. Today, more than 700 lighting industry suppliers have signed up to participate. In announcing the program, the director equates energy effectiveness in lighting with better lighting quality, higher worker productivity, enhanced profitability and reduction of pollution. These are noble and well-defined goals. In logical sequence for goal satisfaction, the engineer should proceed with his project by first recognizing the workers' visual task requirements and then making an appropriate design analysis of the proposed system whether new or retrofitting. Only the appropriate design approach can lead to energy/cost savings, and, more importantly, enhance the workers' visual performance and productivity as well.

The author's extensive pioneering experience in designing energy effective illuminating systems and numerous retrofitting projects has built a solid foundation for the preparation of this book. It is hoped that practicing engineers all over the world will make good use of the helpful information presented in the book. The author believes that a better environment will be created as the result of these collective efforts in designing and implementing energy effective illuminating systems everywhere.

# Chapter 1

# Introduction

## 1.1 Introduction of Illumination

The function of a space can greatly influence the way in which illumination is applied. The typical visual task may be different for an office versus a factory, a store, or a home. Other factors, such as economics, appearance, continuity of effort, and quality of illumination results desired influence the design developed for the task. Thus application techniques generally designated as industrial illumination, store illumination, office illumination and so on have been developed based on lighting solutions for the type of visual tasks encountered in each type of occupancy. Each of these is a synthesis of engineering theory, application techniques, and consumer acceptance in a particular field. It is necessary to relate the design of lighting installation to the particular occupancy of the space it is to serve.

The illuminating design engineer needs to understand the visual sense or how we see. Such knowledge is basic in selecting the actual luminance of the task, its immediate surroundings and anything else in the peripheral field of view. These luminances affect visual comfort and the performance of the task.

Many other factors such as glare, diffusion, reflection, color, luminance ratio, etc. have a significant effect on visibility and the ability to see easily, accurately, and quickly. Poor quality illumination can result in material loss of seeing efficiency and undue fatigue. Therefore a poor quality illumination can hardly be considered energy effective even the light sources, ballasts, and fixtures used in such systems are energy efficient.

## 1.2 Importance of System Energy Effectiveness

Achieving energy effectiveness in an industrial illuminating system or other illumination systems is the key for a successful new installation or retrofit project. Lighting retrofits often provide illuminating engineers an opportunity for improving the illumination quality and enhancing workers visual performance and productivity in a space where poor quality illumination is in existence. In order to achieve these goals along with energy/cost savings, the engineer should be guided by the following steps:

(1) Recognize the workers' visual task requirements and make proper design analysis of a proposed system for replacement.
(2) Select the best suited energy efficient light source and equipment for the proposed system.
(3) Optimize the control techniques and the integration of daylight into a proposed system wherever feasible.

In other words, the engineer must not treat the retrofit project as a simple replacement game and consider its success on energy/cost savings as the only measuring sticks.

## 1.3 Objectives of the Book

Since the objectives of this book are far beyond achieving an energy effective retrofitting installation, it is intended to present the latest concepts in illumination design and useful knowledge, facts, and fundamentals applicable to the industrial illumination to the engineers who are responsible for delivering an energy effective system for their plant operations.

To sum up, a proper design of an illuminating system can provide a solid foundation for achieving energy effectiveness of the system whether it be a new installation or a retrofit project. Therefore the ensuing chapters will cover the new concepts of illumination design, important factors affecting industrial illumination, up-to-date information on energy efficient light sources and equipment, control and daylighting techniques to optimize energy utilization, and finally evaluation of system energy effectiveness. In addition, there is a special chapter on floodlighting design fundamentals and the updated information on lighting and energy standards.

# Chapter 2

# New Concepts in Illumination Design

## 2.1 Determination of Illuminance Levels

Among the many new concepts for lighting design, the first to be discussed is the new method of determining illuminance levels. In the past when illuminating engineers wanted to find the recommended illuminance level for a given task, they would look in the lighting handbook to find a recommended level and then design an illuminating system for the task using the value as a minimum. This procedure provides very little latitude for fine-tuning an illumination design. In the new methods, a more comprehensive investigation of required illuminance is performed according to the following steps:

(1) Instead of a single recommended illuminance value, a category letter is assigned. Table 2-1 shows different category letters for a selected group of industries.

## Table 2-1
## Illuminance Categories for Selected Group of Industries

| Area/Activity | Illuminance Category | Area/Activity | Illuminance Category | Area/Activity | Illuminance Category |
|---|---|---|---|---|---|
| Aircraft maintenance | a | Canning and preserving | | Examining (perching) | I |
| | | Initial grading raw material | | Sponging, decating, winding, | |
| Aircraft manufacturing | a | samples | D | measuring | D |
| | | Tomatoes | E | Piling up and marking | E |
| Assembly | | Color grading and cutting | | Cutting | G |
| Simple | D | rooms | F | Pattern making, preparation of | |
| Moderately difficult | E | Preparation | | trimming, piping, canvas and | |
| Difficult | F | Preliminary sorting | | shoulder pads | E |
| Very difficult | G | Apricots and peaches | D | Fitting, bundling, shading, | |
| Exacting | H | Tomatoes | E | stitching | D |
| | | Olives | F | Shops | F |
| Automobile manufacturing | | Cutting and pitting | E | Inspection | G |
| | | Final sorting | E | Pressing | F |
| Bakeries | | Canning | | Sewing | G |
| Mixing room | D | Continuous-belt canning | E | | |
| Face of shelves | D | Sink canning | E | Control rooms (see Electric generating | |
| Inside of mixing bowl | D | Hand packing | D | stations – interior) | |
| Fermentation room | D | Olives | E | | |
| Make-up room | | Examination of canned samples | F | Corridors (see Service spaces) | |
| Bread | D | Container handling | | | |
| Sweet yeast-raised products | D | Inspection | F | Cotton gin industry | |
| Proofing room | D | Can unscramblers | E | Overhead equipment – separators, | |
| Oven room | D | Labeling and cartoning | D | driers, grid cleaners, stick | |
| Fillings and other ingredients | D | | | machines, conveyers, feeders | |
| Decorating and icing | | Casting (see Foundries) | | and catwalks | D |
| Mechanical | D | | | Gin stand | D |
| Hand | E | Central stations (see Electric generating | | Control console | D |
| Scales and thermometers | D | stations) | | Lint cleaner | D |
| Wrapping | D | | | Bale press | D |
| | | Chemical plants (see Petroleum and | | | |
| Book binding | | chemical plants) | | Dairy farms (see Farms) | |
| Folding, assembling, pasting | D | | | | |
| Cutting, punching, stitching | E | Clay and concrete products | | Dairy products | |
| Embossing and inspection | F | Grinding, filter presses, kiln rooms | C | Fluid milk industry | |
| | | Molding, pressing, cleaning, | | Boiler room | D |
| Breweries | | trimming | D | Bottle storage | D |
| Briew house | D | Enameling | E | Bottle sorting | E |
| Boiling and keg washing | D | Color and glazing – rough work | E | Bottle washers | b |
| Filling (bottles, cans, kegs) | D | Color and glazing – fine work | F | Can washers | D |
| | | | | Cooling equipment | D |
| Candy making | | Cleaning and pressing industry | | Filing: inspection | E |
| Box department | D | Checking and sorting | E | Gauges (on face) | E |
| Chocolate department | | Dry and wet cleaning and | | Laboratories | E |
| Husking, winnowing, fat | | steaming | E | Meter panels (on face) | E |
| extraction, crushing and | | Inspection and spotting | G | Pasteurizers | D |
| refining, feeding | D | Pressing | F | Separators | D |
| Bean cleaning, sorting, dipping, | | Repair and alteration | F | Storage refrigerator | D |
| packing, wrapping | D | | | Tanks, vats | |
| Milling | E | Cloth products | | Light interiors | C |
| Cream making | | Cloth inspection | I | Dark interiors | E |
| Mixing, cooking, molding | E | Cutting | G | Thermometer (on face) | E |
| Gum drops and jellied forms | D | Sewing | G | Weighing room | D |
| Hand decorating | D | Pressing | F | Scales | E |
| Hard candy | | | | Dispatch boards (see Electric generating | |
| Mixing, cooking, molding | D | Clothing manufacture (see Sewn | | stations – interior) | |
| Die cutting and sorting | E | Products) | | | |
| Kiss making and wrapping | E | Receiving, opening, storing, | | Electrical equipment manufacturing | |
| | | shipping | D | Impregnating | D |

| Area/Activity | Illuminance Category | Area/Activity | Illuminance Category | Area/Activity | Illuminance Category |
|---|---|---|---|---|---|
| Insulating: coil winding | E | Milk handling equipment and storage area (milk house or milk room) | | Grinding and chipping | F |
| | | General | C | Inspection | |
| Electric generating stations – interior (see also Nuclear power plants | | Washing area | E | Fine | G |
| Air-conditioning equipment, air preheater and fan floor, | | Bulk tank interior | E | Medium | F |
| | | Loading platform | C | Molding | |
| ash sluicing | B | Feeding area (stall barn feed alley, pens, loose housing | | Medium | F |
| Auxiliaries, pumps, tanks, compressors, gauge area | C | feed are) | C | Large | E |
| Battery rooms | D | Feed storage area – forage | | Pouring | E |
| Boiler platforms | C | Haymow | A | Sorting | E |
| Cable room | B | Hay inspection area | C | Cupola | C |
| Coal handling systems | B | Ladders and stairs | C | Shakeout | D |
| Coal pulverizer | C | Silo | A | | |
| Condensers, deaerator floor, evaporator floor, heater floors | B | Silo room | C | Garages – parking (reference 5) | |
| Control rooms | | Feed storage area – grain and concentrate | | Garages – service | |
| Main control boards | D^C | Grain bin | A | Repairs | E |
| Auxiliary control panels | D^C | Concentrate storage area | B | Active traffic areas | C |
| Operator's station | E^C | Feed processing area | B | Write-up | D |
| Maintenance and wiring areas | D | Livestock housing area (community, maternity, individual calf pens, loose housing holding and resting | | Glass works | |
| Emergency operating lighting | C | | | Mix and furnace rooms, pressing and lehr, glassblowing machines | C |
| Gauge reading | D | areas) | B | Grinding, cutting, silvering | D |
| Hydrocarbon and carbon dioxide manifold are | C | Machine storage area (garage and machine shed) | B | Fine grinding, beveling, polishing | E |
| Laboratory | E | Farm shop area | | Inspection, etching and decorating | F |
| Precipitators | B | Active storage area | B | | |
| Screen house | C | General shop area (machinery repair, rough sawing) | D | Glove manufacturing (see Sewn Products) | |
| Soot or slag blower platform | C | Rough bench and machine work (painting, fine storage, ordinary sheet metal work, welding, medium benchwork) | | | |
| Steam headers and throttles | B | | | Hangars (see Aircraft manufacturing) | |
| Switchgear and motor control centers | D | | D | Hat manufacturing | |
| Telephone and communication equipment rooms | D | Medium bench and machine work (fine woodworking, drill press, metal lathe, grinder) | | Dyeing, stiffening, braiding, cleaning, refining | E |
| Tunnels or galleries, piping and electrical | B | | E | Forming, sizing, pouncing, flanging, finishing, ironing | F |
| Turbine building | | Miscellaneous areas | | Sewing | G |
| Operating floor | D | Farm office (see reference 11 in main text) | | | |
| Below operating floor | C | Restrooms (see Service spaces) | | Inspection | |
| Visitor's gallery | C | Pumphouse | C | Simple | D |
| Water treating area | D | | | Moderately difficult | E |
| | | Farms – poultry (see Poultry Industry) | | Difficult | F |
| Elevators (see Service spaces) | | | | Very difficult | G |
| | | Flour mills | | Exacting | H |
| Explosives manufacturing | | Rolling, sifting, purifying | E | Iron and steel manufacturing | G |
| Hand furnaces, boiling tanks, stationary driers, stationary and gravity crystalizers | D | Packing | D | | |
| | | Product control | F | Laundries | |
| Mechanical furnace, generators and stills, mechanical driers, evaporators, filtration, mechanical crystallizers | | Cleaning, screens, man lifts, aisleways and walkways, bin checking | D | Washing | D |
| | | | | Flat work ironing, weighing, listing, marking | D |
| Tanks for cooking, extractors, percolators, nitrators | D | Forge shops | E | Machine and press finishing, sorting | E |
| | | Foundries | | Find hand ironing | E |
| Farms – dairy | | Annealing (furnaces) | D | | |
| Milking operation area (milking parlor and stall barn) | | Cleaning | D | Leather manufacturing | |
| | | Core making | | Cleaning, tanning and stretching, vats | D |
| General | C | Fine | F | Cutting, fleshing and stuffing | D |
| Cow's udder | D | Medium | E | Finishing and scarfing | E |

| Area/Activity | Illuminance Category | Area/Activity | Illuminance Category | Area/Activity | Illuminance Category |
|---|---|---|---|---|---|
| Leather working | | Paint shops | | Machine storage area (garage | |
| Pressing, winding, glazing | F | Dipping, simple spraying, firing | D | and machine shed) | B |
| Grading, matching, cutting, | | Rubbing, ordinary hand painting | | | |
| scarfing, sewing | G | and finishing art, stencil and | | Printing industries | |
| | | special spraying | D | Type foundries | |
| Locker rooms | C | Fine hand painting and finishing | E | Matrix making, dressing type | E |
| | | Extra-fine hand painting and | | Font assembly – sorting | D |
| Machine shops | | finishing | G | Casting | E |
| Rough bench or machine work | D | | | Printing plants | |
| Medium bench or machine work, | | Paper-box manufacturing | E | Color inspection and appraisal | F |
| ordinary automatic machines, | | | | Machine composition | E |
| rough grinding, medium buffing | | Paper manufacturing | | Composing room | E |
| and polishing | E | Beaters, grinding, calendering | D | Presses | E |
| Fine bench or machine work, fine | | Finishing, cutting, trimming, | | Imposing stones | E |
| automatic machines, medium | | papermaking machines | E | Proofreading | F |
| grinding, fine buffing and | | Hand counting, wet end of paper | | Electrotyping | |
| polishing | G | machine | E | Molding, routing, finishing, leveling | |
| Extra-fine bench or machine work, | | Paper machine reel, paper inspection, | | molds, trimming | E |
| grinding, fine work | H | and laboratories | F | Blocking, tinning | D |
| | | Rewinder | F | Electroplating, washing, | |
| Materials handling | | | | backing | D |
| Wrapping, packing, labeling | D | Petroleum and chemical plants | a | Photoengraving | |
| Picking stock, classifying | D | | | Etching, staging, blocking | D |
| Loading, inside truck bodies | | Plating | D | Routing, finishing, proofing | E |
| and freight cars | C | | | Tint laying, masking | E |
| | | Polishing and burnishing (see Machine | | | |
| Meat packing | | shops) | | Quality Control (see Inspection) | |
| Slaughtering | D | | | | |
| Cleaning, cutting, cooking, grinding, | | Power plants (see Electric generating | | Receiving and shipping (see Materials | |
| canning, packing | D | stations) | | handling) | |
| | | | | | |
| Nuclear power plants (see also Electric | | Poultry industry (see also Farm – dairy) | | Rubber goods – mechanical | a |
| generating stations) | | Brooding, production, and laying | | | |
| Auxiliary building, uncontrolled | | houses | | Rubber tire manufacturing | a |
| access areas | C | Feeding, inspection, cleaning | C | | |
| Controlled access areas | | Charts and records | D | Safety | |
| Count room | E[c] | Thermometers, thermostats, | | | |
| Laboratory | E | time clocks | D | Sawmills | |
| Health physics office | F | Hatcheries | | Secondary log deck | B |
| Medical aid room | F | General area and loading | | Head saw (cutting area viewed by | |
| Hot laundry | D | platform | C | sawyer) | E |
| Storage room | C | Inside incubators | D | Head saw outfeed | B |
| Engineered safety features | | Dubbing station | F | Machine in-feeds (bull edger, resaws, | |
| equipment | D | Sexing | H | edgers, trim, hula saws, | |
| Diesel generator building | D | Egg handling, packing, and shipping | | planers) | B |
| Fuel handling building | | General cleanliness | E | Main mill floor (base lighting) | A |
| Operating floor | D | Egg quality inspection | E | Sorting tables | B |
| Below operating floor | C | Loading platform, egg storage | | Rough lumber grading | D |
| Off gas building | C | area, etc. | C | Finished lumber grading | F |
| Radwaste building | D | Egg processing | | Dry lumber warehouse (planer) | C |
| Reactor building | | General lighting | E | Dry kiln coiling shed | B |
| Operating floor | D | Fowl processing plant | | Chipper infeed | B |
| Below operating floor | C | General (excluding killing and | | Basement areas | |
| | | unloading area) | E | Active | A |
| Offices (see reference 11 in main text) | | Government inspection station | | Inactive | A |
| | | and grading stations | E | Filing room (work areas) | E |
| Packing and boxing (see Materials | | Unloading and killing area | C | | |
| handling) | | Feed storage | | Service spaces (see also Storage rooms) | |
| | | Grain, feed rations | C | Stairways, corridors | B |
| Paint manufacturing | | Processing | C | Elevators, freight and passenger | B |
| Processing | D | Charts and records | D | Toilets and wash rooms | C |
| Mix comparison | F | | | | |

| Area/Activity | Illuminance Category | Area/Activity | Illuminance Category | Area/Activity | Illuminance Category |
|---|---|---|---|---|---|
| **Sewn products** | | setters, sluggers, randers, wheelers, treers, cleaning, spraying, buffing, polishing, embossing | F | **Carding** (nonwoven web formation) | D[e] |
| Receiving, packing, shipping | E | | | Drawing (gilling, pin drafting) | D |
| Opening, raw goods storage | E | **Shoe manufacturing – rubber** | | Combing | D[e] |
| Designing, pattern-drafting, pattern grading and markermaking | F | Washing, coating, mill run compounding | D | Roving (slubbing, fly frame) | E |
| Computerized designing, pattern-making and grading, digitizing, marker-making, and plotting | B | Varnishing, vulcanizing, calendering, upeer and sole cutting | D | Spinning (cab spinning, twisting, texturing) | E |
| Cloth inspection and perching | I | Sole rolling, lining, making and finishing processes | E | **Yarn preparation** | |
| Spreading and cutting (includes computerized cutting) | F[g] | | | Winding, quilling, twisting | E |
| Fitting, sorting and blunding, shading, stitch marking | G | **Soap manufacturing** | | Warping (beaming, sizing) | F[d] |
| Sewing | G | Kettle houses, cutting, soap chip and powder | D | Warp tie-in or drawing-in automatic | E |
| Pressing | F | Stamping, wrapping and packing, filing and packing soap powder | D | **Fabric production** | |
| In-process and final inspection | G | | | Weaving, knitting, tufting | F |
| Finished goods storage and picking orders | F[h] | **Stairways** (see Service spaces) | | Inspection | G[d] |
| Trim preparation, piping, canvas and shoulder pads | F | **Steel** (see Iron and Steel) | | **Finishing** | |
| Machine repair shops | G | | | Fabric preparation (desizing, sourcing, bleaching, singeing, and mercerization) | D |
| Knitting | F | **Storage battery manufacturing** | D | Fabric dyeing (printing) | D |
| Sponging, decating, rewinding, measuring | E | **Storage rooms or warehouses** | | Fabric finishing (calendaring, sonforizing, sueding, chemical treatment) | E[d] |
| Hat manufacture (see Hat Manufacture) | | Inactive | B | Inspection | G[d,f] |
| Leather working (see Leather Working) | | Active | | **Tobacco products** | |
| Shoe manufacturing (see Shoe Manufacturing | | Rough, bulky items | C | Drying, stripping | D |
| | | Small items | D | Grading and sorting | F |
| **Sheet metal works** | | **Structural steel fabrication** | E | **Toilets and wash rooms** (see Service spaces) | |
| Miscellaneous machines, ordinary bench work | E | **Sugar refining** | | | |
| Presses, shears, stamps, spinning, medium bench work | E | Grading | E | **Upholstering** | F |
| Punches | E | Color inspection | F | **Warehouse** (see Storage rooms) | |
| Tin plate inspection, galvanized | F | **Testing** | | **Welding** | |
| Scribing | F | General | D | Orientation | D |
| | | Exacting tests, extra-fine instruments, scales, etc. | F | Precision manual arc-welding | H |
| **Shoe manufacturing – leather** | | **Textile mills** | | **Woodworking** | |
| Cutting and stitching | | Staple fiber preparation | | Rough sawing and bench work | D |
| Cutting tables | G | Stock dyeing, tinting | D | Sizing, planing, rough sanding, medium quality machine and bench work, gluing, veneering, cooperage | D |
| Marking, buttonholing, skiving, sorting, vamping, counting | G | Sorting and grading (wood and cotton | E[d] | Fine bench and machine work, fine sanding and finishing | E |
| Stitching, dark materials | G | Yarn manufacturing | | | |
| Making and finishing, nailers, sole layers, welt beaters and scarfers, trimmers, welters, lasters, edge | | Opening and picking (chute feed | D | | |

a   Industry representatives have established a table of single illuminance values which, in their opinion, can be used in preference to employing reference 6. Illuminance values for specific operations can also be determined using illuminance categories of similar tasks and activities found in this table and the application of the appropriate weighting factors in Table 3.

b   Special lighting such that (1) the luminous area is large enough to cover the surface which is being inspected and (2) the luminance is within the limits necessary to obtain comfortable contrast conditions. This involves the use of sources of large area and relatively low luminance in which the source luminance is the principal factor rather than the illuminance produced at a given point.

c   Maximum levels – controlled system.

d   Supplementary lighting should be provided in this space to produce the higher levels required for specific seeing tasks involved.

e   Additional lighting needs to be provided for maintenance only.

f   Color temperature of the light source is important for color matching.

g   Higher levels from local lighting may be required for manually operated cutting machines.

h   If color matching is critical, use illuminance category G.

(2) The category letters are used to define a range of illuminance. Table 2-2 details illuminance categories and illuminance values for generic types of activities in interiors.

Table **2-2** Illuminance Categories and Illuminance Values for Generic Types of Activities in Interiors

| Type of Activity | Illuminance Category | Ranges of Illuminances | | Reference Work-Plane |
| --- | --- | --- | --- | --- |
| | | Lux | Footcandles | |
| Public spaces with dark surroundings | A | 20–30–50 | 2–3–5 | |
| Simple orientation for short temporary visits | B | 50–75–100 | 5–7.5–10 | General lighting throughout spaces |
| Working spaces where visual tasks are only occasionally performed | C | 100–150–200 | 10–15–20 | |
| Performance of visual tasks of high contrast or large size | D | 200–300–500 | 20–30–50 | |
| Performance of visual tasks of medium contrast of small size | E | 500–750–1000 | 50–75–100 | Illuminance on task |
| Performance of visual tasks of low contrast or very small size | F | 1000–1500–2000 | 100–150–200 | |
| Performance of visual tasks of low contrast and very small size over a prolonged period | G | 2000–3000–5000 | 200–300–500 | |
| Performance of very prolonged and exacting visual tasks | H | 5000–7500–10000 | 500–750–1000 | Illuminance on task, obtained by a combination of general and local (supplementary lighting) |
| Performance of very special visual tasks of extremely low contrast and small size | I | 10000–15000–20000 | 1000–1500–2000 | |

(3) From within the recommended range of illuminance, a specific value of illuminance is selected after consideration is given to the average age of workers, the importance of speed and accuracy, and the reflectance of task background.

The importance of acknowledging the speed and accuracy with which a task must be performed is readily recognized. Less obvious is the need to consider the age of workers and the reflectance of task background. To compensate for reduced visual acuity, more illuminance is needed. Using the average age of workers as the age criterion is a compromise between the need of the young and older workers and, therefore, a valid criterion.

Task background affects the ability to see because it affects contrast, an important aspect of visibility. More illuminance is required to enhance the visibility of tasks with poor contrast. Reflectance is calculated by dividing the reflected value by the incident value. The data given in Tables 2-3 and 2-4 are taken from the IES Lighting Handbook and are applied to provide a single value of illuminance from within the range recommended.

Table **2-3** Weighting Factors for Selecting Specific Illuminance Within Ranges A, B, and C

| Occupant and room characteristics | Weighting factor | | |
|---|---|---|---|
| | −1 | 0 | +1 |
| Workers' age (average) | Under 40 | 40 to 55 | Over 55 |
| Average room reflectance* | >70 percent | 30 to 70 percent | <30 percent |

SOURCE: *IES Lighting Handbook,* sixth edition.

NOTE: This table is used for assessing weighting factors in rooms where a task is not involved.

1. Assign the appropriate weighting factor for each characteristic.

2. Add the two weights; refer to Table 11, Categories A through C:
   a. If the algebraic sum is −1 or −2, use the lowest range value.
   b. If the algebraic sum is 0, use the middle range value.
   c. If the algebraic sum is +1 or +2, use the highest range value.

\* To obtain average room reflectance: determine the areas of ceiling, walls and floor; add the three to establish room surface area; determine the proportion of each surface area to the total; multiply each proportion by the pertinent surface reflectance; and add the three numbers obtained.

Table **2-4**  Weighting Factors for Selecting Specific Illuminance Within
Ranges D through I

| Task or worker characteristics | Weighting factor | | |
|---|---|---|---|
| | **−1** | **0** | **+1** |
| Workers' age (average) | Under 40 | 40 to 55 | Over 55 |
| Speed or accuracy* | Not important | Important | Critical |
| Reflectance of task background, percent | >70 percent | 30 to 70 percent | <30 percent |

SOURCE: *IES Lighting Handbook,* sixth edition.

NOTE: Weighting factors are based upon worker and task information.

1.  Assign the appropriate factor for each characteristic.

2.  Add the three weighting factors and refer to Table 11, Categories D through I:
    a.  If the algebraic sum is –2 or –3, use the lowest range value.
    b.  If the algebraic sum is –1, 0, or +1, use the middle range value.
    c.  If the algebraic sum is +2 or +3, use the highest range value.

 *  Evaluation of speed and accuracy requires that time limitations, the effect of error on safety, quality, and cost, etc., be considered. For example, leisure reading imposes no restrictions on time, and errors are seldom costly or unsafe. Reading engineering drawings or a micrometer requires accuracy and, sometimes, speed. Properly positioning material in a press or mill can impose demands on safety, accuracy, and time.

Illuminating system design can begin after the desired value of illuminance for a given task has been determined. Based on the IES Handbook, the zonal cavity method of determining the number of luminaires and lamps to yield a specified maintained luminance remains unchanged.

## 2.2  Illumination Computational Methods

**2.2.1 Zonal Cavity Method** – Introduced in 1964, the zonal cavity method of performing lighting computations has gained rapid acceptance as the preferred way to calculate number and placement of luminaires required to satisfy a specified illuminance level requirement. Zonal cavity provides a higher degree of accuracy than does the old Lumen Method, because it gives individual consideration to factors that are glossed over empirically in the Lumen Method.

**Definition of Cavities** – With zonal cavity method, the room is considered to contain three vertical zones of cavities. Figure 2-1 defines the various cavities used in this method of computation. Height for luminaire to ceiling is designated as the ceiling cavity (h cc). Distance from luminaire to the work plane is the room cavity (h rc), and the floor cavity (h fc) is measured from the work plane to the floor.

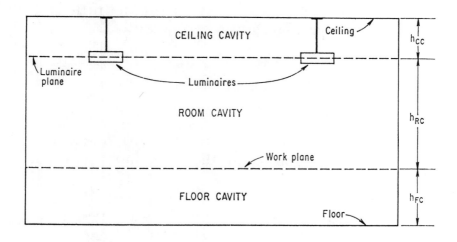

Figure **2-1** Basic cavity divisions of space

To apply the zonal cavity method, it is necessary to determine a parameter known as the "cavity ratio" for each of the three cavities. Following is the formula for determining the cavity ratio:

$$\text{cavity ratio} = \frac{5h \text{ (room length + room width)}}{\text{room length x room width}}$$

where h = h cc for ceiling cavity ratio (CCR)
      = h rc for room cavity ratio (RCR)
      = h fc for floor cavity ratio (FCR)

**2.2.2 Lumen Method Details** – Because of the ease of application of the Lumen Method which yields the average illumination in a room, it is usually employed for larger areas, where the illumination is substantially uniform. The Lumen Method is based on the definition of a footcandle equaling one lumen per square foot.

$$\text{footcandle} = \frac{\text{lumen striking an area}}{\text{square feet of area}}$$

In order to take into consideration such factors as dirt on the luminaire, general depreciation in lumen output of the lamp, and so on, the above formula is modified as follows:

$$\text{footcandle} = \frac{\text{lamps/luminaire x lumens / lp x CU x LLF}}{\text{area / luminaire}}$$

In using the Lumen Method, the following key steps should be taken:

a) To determine the required level of illuminance.
b) To determine the CU (Coefficient of Utilization*) which is the ratio of the lumens reaching the work-

ing plane to the total lumens generated by the lamps. This is a factor that takes into account the efficiency and the distribution of the luminaire, its mounting height, the room proportions, and the reflectances of the walls, ceiling, and floor. Rooms are classified according to shape by 10 room cavity numbers. The cavity ratio can be calculated using the formula given in 2.2.1. The Coefficient of Utilization is selected from tables prepared for various luminaires by manufacturers.

c) To determine the LLF (Light Loss Factor*) – the final Light Loss Factor is the product of all the contributing loss factors. Lamp manufacturers rate filament lamps in accordance with their output when the lamp is new; vapor discharge lamps (fluorescent, mercury, and other types) are rated in accordance with their output after 100 hours of burning.

d) To calculate the number of lamps and luminaires required:

$$\text{no. of lamps} = \frac{\text{footcandles x are}}{\text{lumens / lp x CU x LLF}}$$

$$\text{no. of luminaires} = \frac{\text{no. of lamps}}{\text{lamps / luminaire}}$$

e) To determine the location of the luminaire – luminaire locations depend on the general architecture, size of bays, type of luminaire, position of previous outlets, and so on.

**2.2.3 Point-By-Point Method** – Although currently lighting computations emphasize the Zonal Cavity Method, there is still considerable merit in the Point-By-Point Method. This method lends itself especially well to calculating the illumination level at a particular point where total illumination is the sum of general overhead lighting and supplementary lighting. In this method, information from luminaire candlepower distribution curves must be applied to the mathematical relationship. The total contribution from all luminaires to the illumination level on the task plane must be summed.

a) **Direct illumination component** – The angular coordinate system is most applicable to continuous rows of fluorescent luminaires. Two angles and involved: a longitudinal angle α and a lateral angle β. Angle α is the angle between a vertical line passing through the seeing task to the end of the rows of luminaires. Angle α is easily determined graphically from a chart showing angles α & β for various combinations of V & H. Angle β is the angle between the vertical plane of the row of luminaires and a tilted plane containing both the seeing task and the luminaire or row of luminaires.

   Figure 2-2 shows how angles a & b are defined. The direct illumination component for each luminaire or row of luminaires is determined by referring to the table of direct illumination components for the specific luminaire. The direct illumination components are based on the assumption that the luminaire is mounted 6 ft above the seeing task. If this mounting height is other than 6 ft, the direct illumination component shown in Table 2-5 must be multiplied by 6/V, where V is the mounting height above the task. Thus the total direct illumination component would be the product of 6/V and the sum of the individual direct illumination component of each row.

## Table 2-5  Direct Illumination Components for Category III Luminaire (Based on F40 Lamps Producing 3100 Lumens)

| **Direct Illumination Components** | | | | | | | | | | | | | | | |
|---|---|---|---|---|---|---|---|---|---|---|---|---|---|---|---|
| β | 5 | 15 | 25 | 35 | 45 | 55 | 65 | 75 | 5 | 15 | 25 | 35 | 45 | 55 | 65 | 75 |
| α | **Vertical Surface Illumination Footcandles at a Point On a Plane Parallel to Luminaires** | | | | | | | | **Vertical Surface Illumination Footcandles at a Point On a Plane Perpendicular to Luminaires** | | | | | | | |
| 0-10 | .9 | 2.6 | 3.6 | 3.9 | 3.3 | 1.9 | .7 | .1 | .9 | .8 | .7 | .5 | .3 | .1 | ... | ... |
| 0-20 | 1.8 | 5.0 | 7.0 | 7.7 | 6.6 | 3.8 | 1.5 | .2 | 3.6 | 3.2 | 2.7 | 1.9 | 1.2 | .5 | .1 | ... |
| 0-30 | 2.6 | 7.2 | 10.1 | 11.3 | 9.8 | 5.7 | 2.3 | .3 | 7.7 | 7.0 | 5.8 | 4.3 | 2.7 | 1.1 | .3 | ... |
| 0-40 | 3.2 | 9.0 | 12.8 | 14.5 | 12.9 | 7.7 | 3.2 | .5 | 12.6 | 11.6 | 9.7 | 7.5 | 4.9 | 2.1 | .6 | ... |
| 0-50 | 3.7 | 10.3 | 14.9 | 17.1 | 15.7 | 9.6 | 4.3 | .7 | 17.8 | 16.6 | 14.2 | 11.2 | 7.7 | 3.4 | 1.1 | .1 |
| 0-60 | 4.0 | 11.2 | 16.3 | 18.8 | 17.6 | 11.3 | 5.5 | 1.0 | 22.6 | 21.2 | 18.4 | 14.7 | 10.4 | 5.1 | 1.9 | .2 |
| 0-70 | 4.1 | 11.6 | 17.0 | 19.8 | 18.9 | 12.7 | 6.8 | 1.4 | 26.2 | 24.7 | 21.8 | 17.8 | 13.1 | 7.2 | 3.2 | .3 |
| 0-80 | 4.1 | 11.7 | 17.3 | 20.2 | 19.4 | 13.3 | 7.4 | 1.9 | 28.2 | 26.7 | 23.8 | 19.7 | 14.9 | 8.7 | 4.3 | .8 |
| 0-90 | 4.1 | 11.7 | 17.3 | 20.2 | 19.4 | 13.4 | 7.5 | 2.0 | 28.6 | 27.1 | 24.2 | 20.1 | 15.3 | 9.1 | 4.7 | 1.1 |

| | **F.C. at a Point on Workplane** | | | | | | | |
|---|---|---|---|---|---|---|---|---|
| 0-10 | 10.6 | 9.5 | 7.6 | 5.5 | 3.3 | 1.3 | .3 | ... |
| 0-20 | 20.6 | 18.5 | 14.9 | 10.9 | 6.6 | 2.6 | .7 | ... |
| 0-30 | 29.4 | 26.5 | 21.6 | 16.0 | 9.8 | 4.0 | 1.1 | ... |
| 0-40 | 36.5 | 33.1 | 27.4 | 20.6 | 12.9 | 5.4 | 1.5 | ... |
| 0-50 | 41.8 | 38.1 | 31.9 | 24.3 | 15.7 | 6.7 | 2.0 | .1 |
| 0-60 | 45.2 | 41.3 | 34.8 | 26.8 | 17.6 | 7.9 | 2.6 | .2 |
| 0-70 | 46.9 | 43.0 | 36.4 | 28.3 | 18.9 | 8.9 | 3.2 | .3 |
| 0-80 | 47.4 | 43.6 | 36.9 | 28.8 | 19.4 | 9.3 | 3.5 | .4 |
| 0-90 | 47.5 | 43.7 | 37.0 | 28.8 | 19.4 | 9.3 | 3.5 | .4 |

Category III

2 T-12 Lamps — Any Loading
For T-10 Lamps — C.U. × 1.02

### Luminance Coefficients for 20% Effective Floor Cavity Reflectance

| Ceiling Cavity | | **Reflectances** | | | | | | | | | | | |
|---|---|---|---|---|---|---|---|---|---|---|---|---|---|
| | | 80 | | 50 | | 10 | | 80 | | 50 | | 10 | |
| Walls | | 50 | 30 | 50 | 30 | 50 | 30 | 50 | 30 | 50 | 30 | 50 | 30 |
| WDRC | RCR | **Wall Luminance Coefficients** | | | | | | **Ceiling Cavity Luminance Coefficients** | | | | | |
| .281 | 1 | .246 | .140 | .220 | .126 | .190 | .109 | .230 | .209 | .135 | .124 | .025 | .023 |
| .266 | 2 | .232 | .127 | .209 | .115 | .182 | .102 | .222 | .190 | .130 | .113 | .024 | .021 |
| .245 | 3 | .216 | .115 | .196 | .105 | .172 | .095 | .215 | .176 | .127 | .105 | .024 | .020 |
| .226 | 4 | .202 | .102 | .183 | .097 | .161 | .088 | .209 | .164 | .124 | .099 | .023 | .019 |
| .212 | 5 | .191 | .097 | .173 | .090 | .154 | .082 | .204 | .156 | .121 | .094 | .023 | .018 |
| .196 | 6 | .178 | .090 | .163 | .084 | .145 | .076 | .200 | .149 | .118 | .090 | .022 | .017 |
| .182 | 7 | .168 | .083 | .153 | .078 | .136 | .071 | .194 | .144 | .115 | .087 | .022 | .017 |
| .170 | 8 | .158 | .077 | .145 | .072 | .130 | .066 | .190 | .139 | .113 | .085 | .021 | .016 |
| .159 | 9 | .150 | .072 | .138 | .068 | .123 | .062 | .185 | .135 | .110 | .082 | .021 | .016 |
| .149 | 10 | .141 | .068 | .130 | .064 | .116 | .059 | .180 | .131 | .107 | .080 | .020 | .016 |

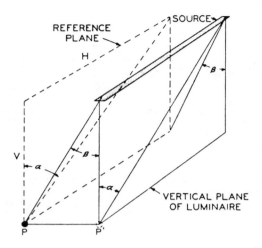

Figure **2-2**   Definition of angular coordinate system for direct
illumination component

b) **Reflected illumination components**
*On the horizontal surface* – This is calculated in
exactly the same manner as the average illumination
using the Lumen Method, except that the RRC
(reflected radiation coefficient) is substituted for the
coefficient of utilization.

$$FC_{RH} = \frac{\text{lamps/luminaire x lumens/lp x RRC x LLF}}{\text{area/luminaire}}$$

where   $RRC = LC_W + RPM (LC_{CC} - LC_W)$

where   $LC_W$ = wall luminance coefficient
$LC_{CC}$ = ceiling cavity luminance coefficient
RPM = room position multiplier

The wall luminance coefficient and the ceiling cavity luminance coefficient are selected for the appropriate room cavity ratio and proper wall and ceiling cavity reflectances from the table of luminance coefficients in the same manner as the coefficient of utilization. The room position multiplier is a function of the room cavity ratio and of the location in the room of the point where the illumination is desired. Table 2-6 lists the value of the RPM for each possible location of the part in the rooms of all room cavity ratios.

Figure 2-3 shows a grid diagram that illustrates the method of designating the location in the room by a letter and a number.

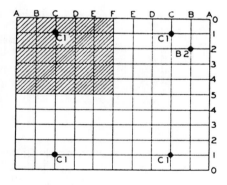

Figure **2-3** – Grid diagram for locating points on the work plane

## Table 2-6 Room position Multipliers

**Room Cavity Ratio = 1**

|   | A | B | C | D | E | F |
|---|---|---|---|---|---|---|
| 0 | .24 | .42 | .47 | .48 | .44 | .48 |
| 1 | .42 | .74 | .81 | .83 | .84 | .84 |
| 2 | .47 | .81 | .90 | .92 | .93 | .93 |
| 3 | .48 | .83 | .92 | .94 | .95 | .95 |
| 4 | .48 | .84 | .93 | .95 | .96 | .97 |
| 5 | .48 | .84 | .93 | .95 | .97 | .97 |

**Room Cavity Ratio = 6**

|   | A | B | C | D | E | F |
|---|---|---|---|---|---|---|
| 0 | .20 | .23 | .26 | .28 | .29 | .30 |
| 1 | .23 | .26 | .29 | .31 | .33 | .36 |
| 2 | .26 | .29 | .35 | .37 | .38 | .40 |
| 3 | .28 | .31 | .37 | .39 | .41 | .43 |
| 4 | .29 | .33 | .38 | .41 | .43 | .45 |
| 5 | .30 | .36 | .40 | .43 | .45 | .47 |

**Room Cavity Ratio = 2**

|   | A | B | C | D | E | F |
|---|---|---|---|---|---|---|
| 0 | .24 | .36 | .42 | .44 | .46 | .46 |
| 1 | .36 | .51 | .60 | .63 | .66 | .68 |
| 2 | .42 | .60 | .68 | .72 | .78 | .83 |
| 3 | .44 | .63 | .72 | .77 | .82 | .85 |
| 4 | .46 | .66 | .78 | .82 | .85 | .86 |
| 5 | .46 | .68 | .83 | .85 | .86 | .87 |

**Room Cavity Ratio = 7**

|   | A | B | C | D | E | F |
|---|---|---|---|---|---|---|
| 0 | .18 | .21 | .23 | .25 | .26 | .27 |
| 1 | .21 | .23 | .26 | .28 | .29 | .30 |
| 2 | .23 | .26 | .30 | .32 | .33 | .34 |
| 3 | .25 | .28 | .32 | .34 | .35 | .36 |
| 4 | .26 | .29 | .33 | .35 | .37 | .37 |
| 5 | .27 | .30 | .34 | .36 | .37 | .38 |

**Room Cavity Ratio = 3**

|   | A | B | C | D | E | F |
|---|---|---|---|---|---|---|
| 0 | .23 | .32 | .37 | .40 | .42 | .42 |
| 1 | .32 | .40 | .48 | .51 | .53 | .57 |
| 2 | .37 | .48 | .58 | .61 | .64 | .67 |
| 3 | .40 | .51 | .61 | .65 | .69 | .71 |
| 4 | .42 | .53 | .64 | .69 | .73 | .75 |
| 5 | .42 | .57 | .67 | .71 | .75 | .77 |

**Room Cavity Ratio = 8**

|   | A | B | C | D | E | F |
|---|---|---|---|---|---|---|
| 0 | .17 | .18 | .21 | .22 | .22 | .23 |
| 1 | .18 | .20 | .23 | .25 | .26 | .26 |
| 2 | .21 | .23 | .26 | .27 | .28 | .29 |
| 3 | .22 | .25 | .27 | .29 | .30 | .30 |
| 4 | .22 | .26 | .28 | .30 | .31 | .32 |
| 5 | .23 | .26 | .29 | .30 | .31 | .32 |

**Room Cavity Ratio = 4**

|   | A | B | C | D | E | F |
|---|---|---|---|---|---|---|
| 0 | .22 | .28 | .32 | .35 | .37 | .37 |
| 1 | .28 | .33 | .40 | .42 | .44 | .48 |
| 2 | .32 | .40 | .48 | .50 | .52 | .57 |
| 3 | .35 | .42 | .50 | .54 | .58 | .61 |
| 4 | .37 | .44 | .52 | .58 | .62 | .64 |
| 5 | .37 | .48 | .57 | .61 | .64 | .66 |

**Room Cavity Ratio = 9**

|   | A | B | C | D | E | F |
|---|---|---|---|---|---|---|
| 0 | .15 | .17 | .18 | .19 | .20 | .20 |
| 1 | .17 | .18 | .20 | .21 | .22 | .23 |
| 2 | .18 | .20 | .23 | .24 | .25 | .25 |
| 3 | .19 | .21 | .24 | .25 | .26 | .26 |
| 4 | .20 | .22 | .25 | .26 | .26 | .27 |
| 5 | .20 | .23 | .25 | .26 | .27 | .27 |

**Room Cavity Ratio = 5**

|   | A | B | C | D | E | F |
|---|---|---|---|---|---|---|
| 0 | .21 | .25 | .28 | .31 | .33 | .33 |
| 1 | .25 | .29 | .33 | .36 | .38 | .42 |
| 2 | .28 | .33 | .40 | .42 | .44 | .48 |
| 3 | .31 | .36 | .42 | .46 | .49 | .52 |
| 4 | .33 | .38 | .44 | .49 | .52 | .54 |
| 5 | .33 | .42 | .48 | .52 | .54 | .56 |

**Room Cavity Ratio = 10**

|   | A | B | C | D | E | F |
|---|---|---|---|---|---|---|
| 0 | .14 | .16 | .16 | .17 | .18 | .18 |
| 1 | .16 | .17 | .18 | .19 | .19 | .20 |
| 2 | .16 | .18 | .19 | .21 | .22 | .22 |
| 3 | .17 | .19 | .21 | .22 | .23 | .23 |
| 4 | .18 | .19 | .22 | .23 | .23 | .24 |
| 5 | .18 | .20 | .22 | .23 | .24 | .25 |

## Table 2-7 Summary of Direct Illumination Components

| Row | $\alpha 1$ | $\alpha 2$ | $\beta$ | Direct illumination component | | Total |
|---|---|---|---|---|---|---|
|  |  |  |  | Front left end | Front right end |  |
| A | 50 | 60[a] | 55 | 6.7 | 7.9 | 14.6 |
| B | 50 | 60 | 25 | 31.9 | 34.8 | 66.7 |
| C | 50 | 60 | 25 | 31.9 | 34.8 | 66.7 |
| D | 50 | 60 | 55 | 6.7 | 7.9 | 14.6 |
|  |  |  |  |  |  | 162.6 |

a Actually $\alpha 2$ is 59, but is rounded off to 60.

*On the vertical surfaces* – To determine illumination reflected to vertical surfaces, the approximate average value is determined using the same general formula, but substituting WRRC (wall reflected radiation coefficient) for the coefficient of utilization:

$$FC_{RV} = \frac{\text{lamps / luminaire x lumens / lp x WRRC x LLF}}{\text{area / luminaire (on work plane)}}$$

where

$$WRRC = \frac{\text{wall luminance coefficient}}{\text{average wall reflectance}} - WDRC$$

where WDRC is the wall direct radiation coefficient which is published for each room cavity ratio together with a table of wall luminance coefficients (see Table 2-5 for a specific type of luminaire).

## 2.2.4 Typical Example

As an example of the calculation of the illumination at a point, assume that four rows of six 4-ft luminaires (for which data are shown in Table 2-5) are surface mounted on 8-ft centers in a room 28 by 30 ft. Assume that the ceiling reflectance (and also that of the ceiling cavity since the luminaires are ceiling mounted) is 80% and that the wall is 50%. Floor cavity reflectance is 20%. The mounting height of the luminaires is 8 1/2 ft above the work plane. The initial illumination on the horizontal work plane at point P is desired. (See Fig. 2-4, a typical luminaire layout plan for this example.)

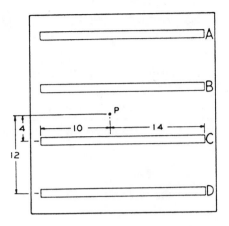

Figure **2-4** A typical luminaire layout plan.

**Calculation of Direct Component** – First let us determine angle α for both ends of the rows of luminaires, and angle β. For angle α, H is 10 ft for $\alpha_1$, and 14 ft for $\alpha_2$. The vertical distance V is 8 1/2 ft. For angle β, H is 12 ft for rows A and D, and is 4 ft for rows B and C. The vertical distance is still 8 1/2 ft. Refer to Table 2-5 for data on the direct illumination component. Table 2-7 summarizes the results of various components as found from the data in Table 2-5. Since the direct illumination component table is all based on a mounting height of 6 ft above the point, and in the case the luminaires are actually 8 1/2 ft above point P, it is necessary to multiply the total fc by 6/8.5. The resultant direct component is 114.8 fc.

**Calculation of Reflected Component** – The RCR for this room is 3.0 and the area per luminaire is 35 square ft. Using the formula for computing the initial value of the reflected illumination component on the horizontal, FC,

$$FC_{RH} = \frac{2 \times 3200 \times \text{reflected radiation coefficient}}{35}$$

The reflected radiation coefficient

$$RRC = LC_W + RPM\ (LC_{CC} - LC_W)$$
$$= 0.216 + 0.75\ (0.215 - 0.216)$$
$$= 0.215$$

RPM is taken from Table 2-6 at point E5, and $FC_{RH} = 39.3$

The total illumination at point P is $114.8 + 39.3 = 154.1$ fc.

## 2.3  Computer Programs for Lighting Design

Several methods of point-by-point computation are available. Such computations can best be handled by computer. Assuming that the necessary programming has been worked out, the computer operator needs only to supply the details of the room, the geometry of the lighting installation, room surface reflectances, and the candlepower data for the luminaire in order to provide a printout of footcandles at the desired points.

### 2.3.1  Computation of ESI Values

Despite the fact that point-by-point computation allows for a particular purpose, this lighting design approach still restricts the illuminating engineers to determining "raw" footcandles and does not provide a measure of whether satisfactory visibility will be produced. With typical reading and writing tasks, visual performance loss caused by veiling reflections should be taken into account. This can only be achieved by equivalent sphere illumination (ESI) computations. The ESI report provided the mathematical details for computing

ESI from candlepower data which are the same as those for the point-by-point computations.

However, the viewing direction of the observer must be specified, as ESI can change substantially with the viewer's orientation. ESI computation is similar to point-by-point computation except the direction of all light rays is analyzed and associated with the reflecting characteristics of the task.

The fundamental qualities involved in the computation of ESI due to a single luminaire are expressed in a form that permits separation of the two variables that give the luminaire's position with respect to an observer. The total effect of all the luminaires in the layout is expressed in a single easily evaluated equation, which is a function of the two variables that give the observer's position. Specification of luminaire placement and orientation is achieved by a simple data input technique. For each of the four viewing directions, the maximum, minimum, average, and mean deviation of ESI values are calculated. These quantities are also calculated for all four directions taken together. The grid point may be as large as 20 by 20, which will give 400 points. This will yield a total 1600 values of ESI. All ESI are printed in an array that corresponds to their positions in the plan view of the room. The value of background luminance (LB), contrast rendition factor (CRF), lighting effectiveness factor (LEF), and effective visibility level are printed in the same format. Figure 2-5 shows a typical room with six points marked, and Table 2-8 gives the resultant sample computer printout.

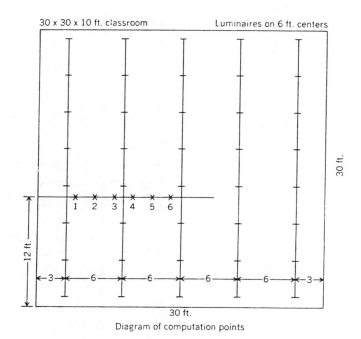

Diagram of computation points

Figure **2-5** Typical 30x30x10 room with six points as marked

Table **2-8** Computer Printout on ESI and Related Computations

|   | Location | | Orientation | | | | | | | |
|---|---|---|---|---|---|---|---|---|---|---|
|   | X | Y | Angle | FC* | FC | LB | LT | CRF | ESI | LEF |
| 1 | 4.00 | 12.00 | 0.0 | 128.87 | 116.26 | 101.28 | 86.54 | .869 | 40.02 | .344 |
|   |   |   | 90.0 |   | 114.95 | 96.17 | 81.35 | .920 | 61.15 | .532 |
| 2 | 6.00 | 12.00 | 0.0 | 139.02 | 135.63 | 116.16 | 97.89 | .939 | 82.02 | .605 |
|   |   |   | 90.0 |   | 122.48 | 105.09 | 89.56 | .882 | 46.61 | .381 |
| 3 | 8.00 | 12.00 | 0.0 | 145.06 | 132.43 | 114.75 | 97.63 | .890 | 53.49 | .404 |
|   |   |   | 90.0 |   | 134.84 | 115.13 | 97.37 | .921 | 70.83 | .525 |
| 4 | 10.00 | 12.00 | 0.0 | 147.29 | 134.71 | 116.61 | 99.17 | .893 | 56.05 | .414 |
|   |   |   | 90.0 |   | 133.86 | 114.68 | 97.19 | .911 | 64.88 | .485 |
| 5 | 12.00 | 12.00 | 0.0 | 147.44 | 144.18 | 123.15 | 103.61 | .947 | 92.16 | .639 |
|   |   |   | 90.0 |   | 131.47 | 113.89 | 96.95 | .888 | 52.03 | .396 |
| 6 | 14.00 | 12.00 | 0.0 | 148.68 | 136.10 | 117.75 | 100.09 | .895 | 57.84 | .425 |
|   |   |   | 90.0 |   | 139.29 | 119.49 | 100.99 | .924 | 74.99 | .538 |

FC* is the footcandle level for no body shadow. All other values include body shadow. LB and LT are the luminances of the background and task respectively.

## 2.3.2  Computer Programs for Lighting Design in General

Computation of ESI values by computer is only one of many computer programs available today.

Lighting design procedures that make full use for the computer programs are still evolving. However, the following effects have proven useful:

(1) In many typical design problems, the locations of the workstations are not known at design time. The iso-ESI plots can be used to help with furniture placement recommendations.

(2) Knowledge of specific workstation locations eliminates some uncertainty and allows the design to be based on the performance calculated at specific points. Calculations at several points in the neighborhood of each of the design locations will reveal the magnitude of change in ESI with change in the observer's position. Knowledge of these deviations is essential for task-oriented lighting design.

(3) The effect of other design parameter uncertainties can be analyzed by recomputing with different parameter values. Examples of this are surface reflectances or lamp lumen output. It allows the design engineers to appraise the effect of possible variations in performance introduced after the design is completed.

The computer can merge the processes of design and drafting by requiring the data needed in these two processes to be entered only once, resulting in a two-way information flow between calculations and graphics. The software is able to interpret graphic information, which exists in its database as numeric data, and insert these numeric data into calculations. The results are then displayed alphanumeric information in tabular or report form. This type of program is commonly known as a computer-aided design and drafting (CADD) system.

A standardized form of presenting luminaire data has been established by the IES and a number of lighting fixture manufacturers. Data needed by the computer to make the calculations are voluminous. As a minimum, lamp data and coefficient of utilization tables, and light loss factors for each luminaire, need to be entered. Availability of these data in an electronic medium is essential in making interactive CADD practical.

### 2.3.3 VCP Values

Many factors are involved in evaluation of the relative comfort of a lighting installation: shape and size of room; reflectances of room surfaces; illumination level; type, size, and light distribution of luminaire used; number and location of luminaires; luminances and their relationship in the entire field of view; location and line of sight of observer; and differences in observer sensitivity to glare. A comprehensive standard evaluation procedure taking all of the foregoing factors into account has come about as the result of numerous extensive investigations. This procedure provides a visual comfort probability (VCP) rating of a given system of lighting. The rating is in terms of the percentage of people who will be expected to find the given lighting system acceptable when they are seated in the most undesirable location.

By means of several procedures outlined in the IES Lighting Handbook, it is possible and useful to study proposed lighting system designs from the standpoint of visual comfort probability (VCP) by preparing tables such as Table 2-9.

Based on the recent software survey by IES, several programs are available for calculating VCP for indoor projects.

### Table **2-9**  Typical VCP Values

Wall reflectance: 50%
Ceiling cavity reflectance: 80%
Effective floor cavity reflectance: 20%

Work plane illumination: 100fc
Luminaire no. 00
Room size: 60 x 30 ft

| Mounting height above the floor (ft) | Luminaires | |
|:---:|:---:|:---:|
| | **Lengthwise** | **Crosswise** |
| 8.5 | 68 | 67 |
| 10 | 69 | 69 |
| 13 | 71 | 70 |
| 16 | 74 | 73 |

## Bibliography

Chen, Kao, New Concepts in Interior Lighting Design, IEEE Transactions on Industry Applications, Sept./ Oct. 1984, pp. 1179-1184.

DiLaura, D.L., Whatever Happened to Equivalent Sphere Illumination, Lighting Design and Applications, Nov. 1982, pp. 17-18.

IES Computer Committee Report, Available Lighting Computer Programs, Lighting Design and Applications, Sept. 1986.

IES Software Survey, Lighting Design and Applications, March 1991, pp. 18-24

Lighting Handbook, Westinghouse Electric Corporation, Bloomfield, N.J., May 1978.

Rowe, G. D., Determining Illumination Requirements, Plant Engineering, Feb. 4, 1982, pp. 69-72.

Sisson, William, Determining Cavity Ratios for Zonal Cavity Lighting Calculations, Plant Engineering, Nov. 27, 1970, pp. 68-69.

# Chapter 3

# Factors Affecting Industrial Illumination

## 3.1 Basic Definitions

**Illuminance** – Illuminance is the density of luminous flux on a surface, expressed in either footcandles (lumens/sq. ft) or lux (lx), (lux = 0.0929 fc).

**Luminance (or photometric Brightness)** – Luminance is the luminous intensity of a surface in a given direction per unit of projected area of the surfaces, expressed in candelas per unit area or in lumens per unit area.

**Reflectance** – Reflectance is the ratio of the light reflected from a surface to that incident upon it. Reflection may be of several types, the most common being specular, diffuse, spread, and mixed.

**Glare** – Glare is any brightness that causes discomfort, interference with vision, or eye fatigue.

**Color Rendering Index\* (CRI)** – in 1964 the CIE (Commission Internationale de l'Eclairage) officially adopted the IES procedure for rating lighting sources and developed the current standards by which light sources are rated for their color rendering properties.

The CRI is a numerical value to the color comparison of one light source to that of a reference light source.

**Color Preference Index\* (CPI)** – The CPI is determined by a similar procedure to that used for the CRI. The difference is that CPI recognizes the very real human ingredient of preference. This index is based on peoples' preference for the coloration of certain identifiable objects, such as complexions, meat, vegetables, fruits, and foliage, to be slightly different then their colors are in daylight. CPI indicates how a source will render colors with respect to how we best appreciate and remember that color.

**Equivalent Sphere Illumination (ESI)** – ESI is a means of determining how well a lighting system will provide task visibility in a given situation. ESI may be predicted for many points in a lighting system through the use of any of several available computer programs; or measured in an installation with any of several different types of meters.

**Visual Comfort Probability (VCP)** – Discomfort glare is most often produced by direct glare from luminances that are excessively bright. Discomfort glare can also be caused by reflected glare, which should not be confused with veiling reflections, which cause a reduction in visual performance rather than discomfort. VCP is based in terms of the percentage of people who will be expected to find the given lighting system acceptable when they are seated in the most undesirable location.

**Coefficient of Utilization (CU)** – CU is the ratio of the lumens reaching the work plane (assumed to be a horizontal plane 30 in. above the floor) to the total lumens generated by the light source. This is a factor that takes into account the efficiency and distribution of the luminaire, its mounting height, the room proportions, and the reflectance of the wall, ceiling, and floor.

**Light Loss Factor (LLF)** – The final LLF is the product of all the contributing loss factors. It is the ratio of the illumination when it reaches its lowest level at the task just before corrective action is taken, to the initial level if none of the contributing loss factors were considered. There are eight contributing loss factors that require consideration:

a) Ballast performance (Ballast Factor)
b) Voltage to luminaires
c) Luminaire reflectance and transmittance changes
d) Lamp outages
e) Luminaire ambient temperature
f) Heat-exchange luminaires
g) Lamp lumen depreciation (LLD)
h) Luminaire dirt depreciation (LDD)

**Lamp Lumen Depreciation Factor (LLD)** – This factor is used in illumination calculations to quantify the output of light source at 70% of their rated life as a percentage of their initial output.

**Ballast Factor** – The ratio between the light output from a lamp operating on a commercial ballast and the light output from the same lamp operating on a reference ballast.

### 3.2 Factors and Remedies

Quality of illumination pertains to the distribution of luminaires in the visual environment. The terms is used in a positive sense and implies that all luminaires contribute favorably to visual performance. However, glare, diffusion, reflection, uniformity, color, luminance and luminance ratio* all have a significant effect on visibility and the ability to see easily, accurately, and quickly. Industrial installations of poor quality are easily recognized as uncomfortable and possibly hazardous. Some of the factors are discussed in more detail below:

**Direct Glare** – When glare is caused by the source of lighting within the field of view, whether daylight or electric, it is defined as direct glare. To reduce direct glare, the following suggestions may be useful:

a) To decrease the brightness of light sources or lighting equipment, or both.
b) To reduce the area of high luminance causing the glare condition.
c) To increase the angle between the glare source and the line of vision.
d) To increase the luminance of the area surrounding the glare source and against which it is seen.

To reduce direct glare, luminaires should be mounted as far above the normal line of sight as possible and should be designed to limit both the luminance and the quality of light emitted in the 45-85 degree zone because such light may interfere with vision. This precaution includes the use of supplementary lighting equipment. There is such a wide divergence of tasks and environmental conditions that it may not be possible to recommend a degree of quality satisfactory to all needs. In production areas, luminaires within the normal field of view should be shielded to at least 25 degrees from the horizontal, preferably to 45 degrees.

**Reflected Glare** – Reflected glare is caused by the reflection of high luminance light sources from shiny surfaces. In the manufacturing area, this may be a particularly serious problem where critical seeing is involved with highly polished sheet metal, vernier scales, and machined metal surfaces. There are several ways to minimize or eliminate reflected glare:

a) Use a light source of low luminance, consistent with the type of work in process and the surroundings.

b) If the luminance of the light source cannot be reduced to a desirable level, it may be possible to orient the work so that reflections are not directed in the normal line of vision.

c) Increasing the level of illumination by increasing the number of sources will reduce the effect of reflected glare by reducing the proportion of illumination provided on the task by sources located in positions causing reflections.

d) In special cases, it may be practical to reduce the specular reflection by changing the specular character of the offending surface.

**Distribution, Reflection, and Shadows** – Uniform horizontal illuminance (maximum and minimum not more than one-sixth above or below the average level) is usually desirable for industrial interiors to permit flexible arrangements of operations and equipment, and to assure more uniform luminance in the entire area.

Reflections of light sources in the task can be useful provided that the reflection does not create reflected glare. In the machining and inspection of small metal parts, reflections can indicate faults in contours, make scribe marks more visible, and so on.

Shadows from the general illumination systems can be desirable for accenting the depth and forms of various objects, but harsh shadows should be avoided. Shadows are softer and less pronounced when large diffusing luminaires are used or the object is illuminated from many sources. Clearly defined shadows are distinct aids in some specialized operations, such as engraving on polished surfaces, some type of bench layout work, or certain textile inspections. This type of shadow effect can best be obtained by supplementary directional lighting combined with ample diffused general illumination.

**Luminance and Luminance Ratios\*** – The ability to see details depends on the contrast between the detail and its background. The greater the contrast difference in luminance, the more readily the seeing task is performed. The eye functions most comfortably and efficiently when the luminance within the remainder of the environment is relatively uniform. In manufacturing, there are many areas where it is not practical to achieve the same luminance relationships as easily as in offices. Table 3-1 is shown as a practical guide to recommended maximum luminance ratios for industrial areas. To achieve the recommended luminance relationships, it is necessary to select the reflectances of all the finishes of the room surfaces and equipment as well as control of the luminance distribution of the lighting equipment. Table 3-2 lists the recommended reflectance values of industrial interiors and equipment. High-reflectance surfaces are desirable to provide the recommended luminance relationships and high utilization of light.

**Color Quality of Light** – In general, for seeing tasks in industrial areas, there appears to be no effect upon visual acuity by variation in color of light. However, where color discrimination or color matching are a part of the work process, such as in the printing and textile industries, the color of light should be carefully selected. Color always has an effect on the appearance of the work space and on the complexions of people. The illuminating system and the decorative scheme should be properly coordinated.

## Table **3-1** Recommended Maximum Luminance Ratios for Industrial Areas

| | Environmental Classification | | |
| --- | --- | --- | --- |
| | A | B | C |
| (1) Between tasks and adjacent darker surroundings | 3 to 1 | 3 to 1 | 5 to 1 |
| (2) Between tasks and adjacent lighter surroundings | 1 to 3 | 1 to 3 | 1 to 5 |
| (3) Between tasks and more remote darker surfaces | 10 to 1 | 20 to 1 | * |
| (4) Between tasks and more remote lighter surfaces | 1 to 10 | 1 to 20 | * |
| (5) Between luminaires (or windows, skylights, etc.) and surfaces adjacent to them | 20 to 1 | * | * |
| (6) Anywhere within normal field of view | 40 to 1 | * | * |

* Luminance ratio control not practical.
A—Interior areas where reflectances of entire space can be controlled in line with recommendations for optimum seeing conditions.
B—Areas where reflectances of immediate work area can be controlled, but control of remote surround is limited.
C—Areas (indoor and outdoor) where it is completely impractical to control reflectances and difficult to alter environmental conditions.

## Table **3-2** Recommended Reflectance Values for Industrial Interiors and Equipment

| Surfaces | Reflectance* (percent) |
| --- | --- |
| Ceiling | 80 to 90 |
| Walls | 40 to 60 |
| Desk and bench tops, machines and equipment | 25 to 45 |
| Floors | not less than 20 |

* Reflectance should be maintained as near as practical to recommended values.

**Veiling Reflections*** – Figure 3-1 shows that light would reflect into eyes of viewer from the "offending zone" and defines the zone of veiling reflection. Veiling reflection would diminish visibility, but the viewer would be unaware of it. the Contrast Rendition Factor* (CRF) can be applied as a measure of the amount of veiling reflection.

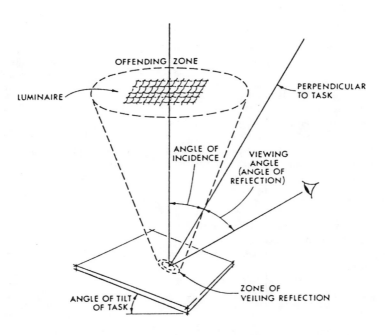

Figure **3-1**  Diagram showing "offending zone" and zone of
veiling reflection

Another important factor is the **Lighting Effective-ness Factor\* (LEF)** – An overall lighting system efficiency factor considers both the quality of light as reference to equivalent sphere illumination and the effects of veiling reflections. Light patterns such as "batwing" can help solve veiling reflection problems. Figure 3-2 shows the light distribution curve of a typical batwing luminaire.

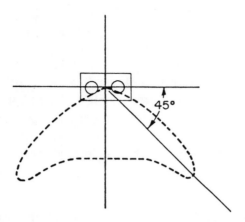

Figure **3-2** A typical "batwing" light distribution

## 3.3 Daylighting

The daylight contribution should be carefully evaluated and should always be coordinated with a planned electric lighting system.

**Fenestration*** – Fenestration has at least three useful purposes in industrial buildings:

a) For the admission, control, and distribution of daylight.
b) For a distant focus for the eyes, which relaxes the eye muscles.
c) To elminate the dissatisfaction many people experience in completely closed-in areas.

An adequate electric lighting system should always be provided because of the wide variation in daylight.

**Building Orientation** – All fenestration should be equipped with control device appropriate to any luminance problems. Special attention should be given to glare control for latitudes where fenestration frequently receives direct sunlight. Diffuse-glaring fixed or adjustable louvers are some of the control means that may be applied.

For an industrial building, windows in the sidewalls admit daylight and natural ventilation, and afford occupants a view out. However, their uncontrolled luminance may be a problem. There are many control means to make the daylight useful to workers' seeing tasks, resulting in energy savings as the ultimate goal.

## Bibliography

Allphin, Willard, Minimizing Veiling Reflections in Lighting Installations, Plant Engineerings, June 1, 1972, pp. 62-63.

Chen, Kao, Industrial Power Distributions and Illuminating Systems, Marcel Dekker, Inc., New York, N.Y., 1990.

Lighting Handbook, Application Volume, Illuminating Engineering Society, New York, N.Y., 1987.

# Chapter 4

# System Components

## 4.1 Light Sources

Incandescent, fluorescent, and/or high-intensity-discharge (HID) lamps are used in industrial lighting. They differ considerably in physical dimensions, electrical characteristics, spectral power distribution, and operating performance. Some are better suited than others to certain applications; however, sometimes two or more sources may qualify to fulfill a specific lighting requirement.

### 4.1.1 Incandescent Filament Lamps

Initial efficacy of typical incandescent lamps (25 to 1000 W) ranges from 10 to 23 lumens per watt. They are commonly designed for approximately 1000 h of life. It is generally known that incandescent lamps conform to the supply voltage. A change of only a few volts can seriously affect both life and light output. There are special types of incandescent lamps:

(1) **Reflectorized (R, PAR, and ER) lamps.**
    These lamps have self-contained reflectors and are manufactured in a number of sizes, from 30 to 1500 W, and in various light distributions. These lamps in

general have a better-maintained illuminance. ER lamps (50, 75, and 120 W) control their beams such that they focus about 2 in. in front of their faces. Especially useful in "baffled downlight" luminaires, they permit high light utilization with attendant savings in energy. Fig. 4-1 shows an ER30 lamp and its beam focus.

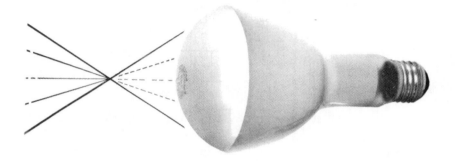

Figure **4-1** Am ER30 lamp and its beam focus

## (2) Special lamps for industrial applications.

a) *Rough service* – Rough service lamps (from 25 to 500 W) are made with extra filament supports to withstand mechanical shock, and are used principally with extension cords.

b) *Silicone-rubber coated lamps* – These lamps have special rubber-like coatings that serve to reduce breakage from both thermal and mechanical shock; or should breakage occur, the glass fragments nearly always remain intact. Available in sizes from 25 to 200 W, they are especially suited to food-packaging industries and to others where manufacturing functions may subject lamps to mechanical damage.

c) *Extended service lamps* – These lamps operate for approximately two to three times the normal rated life. They are useful where cost of lamp replacement is high and cost of power is low.

d) *Thermal shock resistant lamps* – These are available in various wattages and bulb shapes and are recommended for applications where moisture may fracture the hot tub.

## (3) **Spotlight, floodlight and projector lamps.**

Characteristic features of all lamps designed for spotlight, floodlight and projection applications are compact filaments accurately positioned with respect to the base, for the purpose of light control; relatively short life, for high efficacy and luminance; comparatively small bulbs and restricted burning position. some projection lamps for use in certain types of projectors have an opaque coating on the top of the bulb to prevent the emission of stray light.

Low voltage PAR lamps designed at 5.5 and 12 volts produce a concentrated beam pattern with good optical control. They can be used in the fixture which may be recessed, surface or track mounted.

## (4) **Tungsten-halogen lamps.**

These employ halogens to preclude blackening of the tubular envelope. They have extremely good lumen maintenance over a life of 2000 h or more. The shape of the lamps enables the luminaire to provide excellent beam control.

These lamps are both single and double ended. They may also be sealed into outer bulbs such as PAR types for good optical control.

Because of their characteristics, tungsten-halogen lamps find extensive application in floodlighting, aviation, photographic, photocopy, special effects and special application lighting. they have also found broad use in automobile headlighting.

(5) **Infrared lamps.**
Infrared lamps used in the home and for therapeutic purposes are commonly of the convenient self-contained 250 W R40 bulb with internal reflector and red bulb face. Those used in industrial processes are of three types: reflector, clear G-30 bulb, and T3 quartz bulb. Gold plated or specular aluminum reflectors are most effective for use with unreflectorized infrared lamps.

(6) **High- and low-voltage lamps.**
These are available in 100 to 1500 W for 230- and 250 V circuits. They have less rugged filament, require more supports, and are less efficient than are 120 V lamps of equal wattage.

Lamps for operation on 30- and 60 V circuits are also available for use in train lighting and in country home service.

(7) **Energy efficient incandescent lamps.**
Energy-saving potential exists for incandescent lamps where reflector lamps can be used. A new line of indoor reflector lamps (ER) allows reduction of 50% or more in energy consumption in many installations.

In addition, incandescent lamps with special radiation covering envelopes are being developed. Compact high-emissivity filaments are required to absorb infrared radiation reflected from specially coated envelopes. The energy-saving potential for the new lamps could be as much as 60% compared with equivalent conventional incandescent lamps.

More recent developments are the application of a tungsten-halogen cycle and the use of selective reflecting thin film. As already discussed in (4) how the tungsten-halogen cycle works, the selective reflecting thin film comprises of many layers of a particular thickness of the order of the wavelength of light. The film is transparent to visible light and highly reflective to the near and far infrared radiation (IR) greater than 7000 A. The bulb wall is shaped such that the IR is reflected back onto the filament, further heating the filament. This technique increases the lamp efficacy by a factor approaching two.

### 4.1.2 Fluorescent Lamps

The fluorescent lamp is an electric discharge source in which light is produced by the fluorescence of phosphors activated by ultraviolet energy from a low-pressure mercury arc. The lamp requires a ballast to limit the current and, in many instances, to transform the supply voltage. Lamp performance is influenced by the character of the ballast and luminaire, line voltage, ambient temperature, burning hours per start, and air movement. Fluorescent lamps are available in many variations of "white" and in a number of colors. Standard "cool white" is most popular for industrial lighting. The efficacy of cool white lamps varies between 30 and 100 lumens per watt (exclusive of 20% power loss in the ballast). Although most fluorescent lamps have tubular envelopes, there are special types, such as circular, U-shaped, reflectorized, and jacketed.

**Energy Efficient Fluorescent Lamps** – Since the 1970's there has been a line of reduced wattage replacements for stand fluorescent lamps. These lamps are now available in all popular sizes and colors for most applications. Limitations of energy saving reduced-wattage lamps are:

(1) Should be used only where ambient temperature does not drop below 60° F
(2) Should be used only on high power factor ballasts
(3) Not to be used where drafts of cold-air ducts would be directed

Typical energy savings are about 6 W per lamp for the popular 4-ft 40 W replacement and 15 W per lamp for the popular 8-ft slimline 75 W replacement. Savings in energy cost normally pay back the new lamp cost in a year at typical power rates and lamp costs.

**Compact Fluorescent Lamps (CFL)** – A recently developed arc-discharge lamp, called compact fluorescent lamp, can replace an incandescent light source in particular applications. Three configurations are possible for the installation of compact fluorescent lamps: dedicated, self-ballasted, and modular. Dedicated compact fluorescent lamp systems are similar to full-size fluorescent lighting systems in which a ballast is hard-wired to lamp holders within a luminaire. Self-ballasted and modular compact fluorescent lamp products have screwbases designed for installation in medium screw base sockets; they typically replace incandescent lamps. A self-ballasted compact fluorescent lamp contains a lamp and ballast as an inseparable unit. A modular compact fluorescent lamp product consists of a screwbase ballast with a replaceable lamp. The ballast and lamp connect together using a socket-and-base design that ensures compatibility of lamps and ballasts. While most of the modular types are operated in the preheat mode, the electronic ballasted lamps are operated in the rapid start mode and in principle could be dimmed.

Compact fluorescent lamps are developed as an economical substitution for lower wattage incandescent lamps in applications requiring long burning hours, such as corridors, stairwells, lobbies, and reception areas. The economics are two-fold: the CFL has a rated life of about 10,000 hours and even when ballast losses are included, the CFL offers four times the efficacy of an incandescent lamp. In addition, being the first lamp type to use the triphosphor coating, the CFL has a CRI of 80 or better and a 2790° K color temperature, making it compatible in appearance with the chromaticity of an incandescent lamp.

Compact fluorescent lamps have either a T-4 (10mm) or a T-5 (15mm) glass envelope that is bent in a U-shape and mounted on a special base. Ratings are 5, 7, 9, 13, 18 and 26 W. The single twin-tube CFL are too long for some applications, so the family of double twin-tube or "quad" tube lamps was introduced. In many cases, the CFL should be matched to a specific ballast type from a particular manufacturer. If a CFL is compatible with a particular ballast, it may not properly start and operate. Fig. 4-2 shows a modular CFL which is a double-folded, bent-tube assembly that can be retrofitted to a standard medium-base socket.

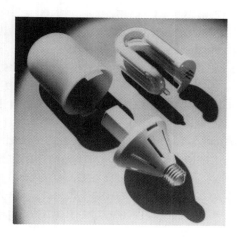

Figure **4-2** A modular compact fluorescent lamp assembly

Recent significant improvements in fluorescent lighting are:

(1) the cathode cutout lamps;
(2) 32W F32T8 lamps

The cathode cutout lamp has a thermal cutout switch in series with the filament, the switch opens the circuit that applies the low voltage (3.6 V) across the electrode after the discharge has been ignited. The efficacy of the lamp is increased by 6% comparing with the standard 40W F40 rapid start lamp. The lamp is not recommended for use with electronic ballasts. The 32W F32T8 lamp is rapidly replacing the standard 40W T12 lamp. It uses the rare earth 70 CRI phosphor and has a lamp efficacy of 90 lm/w at 60 Hz. When operated at higher frequency in the instant mode, its efficacy approaches 100 lm/w, but its life is rated at 15,000 hours.

### 4.1.3  High Intensity Discharge (HID) Lamps

High intensity discharge lamps are electric discharge sources. The basic difference from fluorescent lamps is that HID lamps operate at a much higher arc pressure. Spectral characteristics differ from those of fluorescent lamps because the higher pressure arc emits a large portion of its visible light. HID lamps produce full light output only at full operating pressure usually several minutes after starting. Most HID lamps contain both an inner and an outer bulb. The inner bulb is made of quartz or polycrystalline aluminum; the outer bulb is generally of thermal shock-resistant glass. HID lamps require current-limiting devices, which consume 10 to 20% additional watts. HID lamps include mercury, metal halide, high-pressure sodium, and low-pressure sodium lamps:

(1) **Mercury Lamps** – These are low in efficacy compared to other HID sources, and are obsolescent for most

industrial lighting applications. They are available with either "clear" or phosphor-coated bulbs of 40 to 1000 W, and in various sizes and shapes. Typical efficacy ranges from 30 to 63 lumens per watt, not including ballast loss. "Clear" mercury lamps produce light rich in yellow and green tones while lacking in red. Phosphor-coated lamps provide improved color. Special types include semireflector, reflectorized, and self-ballasted lamps.

(2) **Metal Halide (MH) Lamps** – These are similar in construction to mercury lamps. The difference is in the arc tube, which contains various metal halides in addition to mercury. They are available in either clear or phosphor-coated bulbs from 32 to 1500 W. Present efficacies range from 70 to 125 lumens per watt, not including ballast power loss. Color improvement is achieved by the metal halide additives.

(3) **High Pressure Sodium (HPS) Lamps** – Light is produced by electricity passing through sodium vapor. They are presently available in sizes of 35 to 1000 W. Typical initial efficacies are about twice that of mercury lamps: from 80 to 140 lumens per watt, not including ballast power loss. Normally with clear outer envelopes, they may also be obtained with coatings that improve diffusion. The color of light produced is golden white. Fig. 4-3 shows several 250 W HPS lamps.

(4) **Low Pressure Sodium Lamps** – These are presently available in 35 to 180 W. Typical initial efficacies are high: 137 to 183 lumens per watt, exclusive of ballast power loss. Applications are limited by virtue of their monochromatic yellow color. Fig. 4-4 shows a typical low-pressure sodium lamp.

Figure **4-3** 250 W high pressure sodium lamps

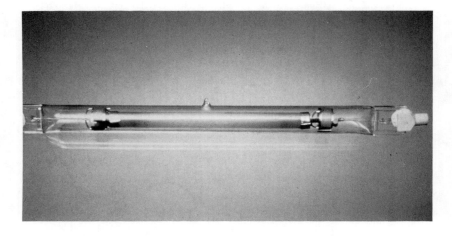

Figure **4-4** Typical low-pressure sodium lamp

## 4.2 Ballasts

### 4.2.1 Fluorescent Ballasts

Most fluorescent lamps operate on one of three types of ballast circuit: preheat, instant start, or rapid start (Figures 4-5 to 4-7). A few can be operated with either preheat or rapid-start ballasts. Preheat lamps up to 20 W can be operated on special rapid-start (trigger-start) ballasts. Preheat lamps operated on preheat ballasts require auxiliary "starters" to allow current to flow through the electrodes for a few moments before the arc is established across the length of the lamps. Instant start and slimline lamps require no starters. The ballasts provide enough voltage to light the lamps instantly. Rapid-start lamp operation is also starterless. Rapid-start lamps are most popular for new fluorescent lighting installations. They are available for ballasts that provide 430-, 800-, 100-, and 1500 mA loadings. Fluorescent lamp ballasts are available for most secondary distribution voltages.

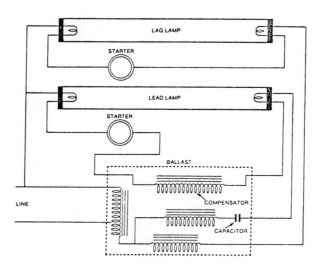

Figure **4-5** Preheat fluorescent ballast circuit

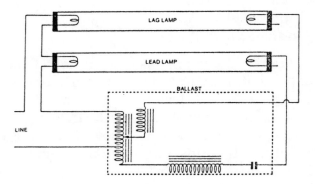

Figure **4-6**  Instant-start fluorescent ballast circuit

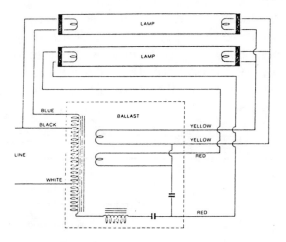

Figure **4-7**  Rapid-start fluorescent ballast circuit

(1) **Energy Efficient Ballasts** – In general, the greater
the ballast size (power rating) the greater is the ballast efficiency. That is, the relative ballast losses are
less for a 40 W ballast than a 20 W ballast. A two-
lamp ballast is more efficient than a one-lamp bal-

last. The ballast efficiencies can generally be calculated from the manufacturers' catalog data. However, the designer shall be certain the manufacturer specifies the test conditions at which the ballast is rated. The internal ballast losses are determined by coil construction, the nature of the magnetic materials, and resistance of the conducting coil wire. With increased energy costs, ballast manufacturers have introduced energy efficient ballasts that minimize the ballast losses. The following sections detail some of the new ballasts:

a) *Low-energy Ballasts* – Fluorescent lamps may be operated at less than rated power and output, provided starting voltage and operating voltage requirements are met. In the case of rapid start lamps, cathodes shall be heated regardless of the current through the lamp to provide rated lamp life. Low Energy ballasts are low-current designs which can provide energy reductions compared to standard units. They are useful where certain luminaire spacing to mounting height criteria shall be followed, and the desired illumination level is less than that obtained by full output operation of the lamp. Before operating fluorescent lamps at less than rated output, it is wise to check the ambient temperatures of the air surrounding the lamps. Higher minimum ambient are required when low energy lamps or standard lamps are used with low-energy ballasts.

b) *High/Low Ballasts* – This type of ballast, which is generally available only for rapid start circuit operation, contains extra leads which can be connected or switched to provide multi-level operation of the lamp. Two-level and three-level rapid start ballasts are available, and fixture output

may be set according to the lighting requirements
of the area. Operation of fluorescent lamps at less
than rated output may raise minimum operating
temperature requirements.

c)  *Low-Loss Ballasts* – Ballasts may be designed to
reduce internal losses by improving mechanical
and electrical characteristics. More efficient magnetic circuits, closer spacing of coils, and improved insulation systems can result in loss
reductions of approximately 50%, compared to
conventional units. Again, the ballast manufacturers' ratings should be consulted to determine
which ballast will have the least losses for the
power system and lamp combination involved.

(2) **Electronic Ballasts** – The efficiency of fluorescent
lamp systems can be improved even more by utilizing
electronic ballasts. Improvements occur two ways:

a)  By way of lower internal losses within the ballasts
b)  By making use of the fact that fluorescent lamp
efficacy increases as a function of the frequency
of the applied power.

Electronic ballasts usually operate lamp at fre-
quencies above 20kHz. In general operating a fluorescent lamp with an electronic ballast will realize a 20
to 25% increase in system efficiency compared to a
standard core-coil ballast. The electronic ballast
offers improved performance that includes: 1) reduced flicker, 2) improved voltage and thermal regulation, and 3) the ability to dim fluorescent lamp over
a wide range without affecting lamp life. Table 4-1
shows the performance of the standard and energy-
efficient core-coil ballasts and also the electronic high
frequency ballasts.

The data are for two lamp systems. The 40W F40T12 cool white lamp is still in general use; however, the 32W F32T8 lamp is gaining in popularity based on its high system efficiency. The table shows that the electronic ballast can meet all necessary lamp parameters to maintain lamp life. The high harmonic data for the T8 electronic ballast was designed before harmonics became an issue. Today the technology is available to reduce the harmonic contents within a specified limit, the constraint would be the cost or economic considerations.

Table **4-1** Performance data of the Standard and Energy-Efficient Ballasts vs. Electronic Ballasts on Fluorescent Lighting Systems

| BALLAST LAMP | EE mag. 40W F40 T-12 CW | EE mag. 40W F40 T-12 CW | Electronic Ballasts | | |
|---|---|---|---|---|---|
| | | | 40W F40 T-12 CW | 32W F32 T-8 41 K | 32W F32 T-8 41 K |
| Power (W) | 85 | 81 | 74 | 69 | 65 |
| Filament Voltage (V) | 3.5 | 0 | 2.6 | 3.4 | 0 |
| Lamp Current Crest Factor | 1.7 | 1.7 | 1.4 | 1.5 | 1.5 |
| Ballast Factor (%) | 93 | 93 | 93 | 100 | 100 |
| Light Output (lm) | 5690 | 5680 | 5670 | 5820 | 5800 |
| Flicker (%) | 30 | 30 | 1 | 1 | 1 |
| 3rd Harmonic (%) | 12 | 12 | 5 | 43 | 43 |
| Ballast Efficiency (%) | 87 | 87 | 89 | 90 | 90 |
| Lamp Efficacy (lm/W) | 77 | 80 | 86 | 93 | 100 |
| System Efficacy (lm/W) | 67 | 70 | 77 | 84 | 90 |

Electronic ballasts are available for all of the commonly used fluorescent lamps, including the 40W F40T12, the 34W F40T12, 40W F40T10, and 32W F32T8 (rapid and instant start), all of the F96 type lamps, and the high output F96 lamps. There are one, two, three, and four lamp ballasts. The new generation of electronic ballasts for CFL has power factor above 90% with harmonics below 20%.

### 4.2.2 High Intensity Discharge Ballasts

Ballast factors for HID lamps are usually close to unity, and ballast circuitry is somewhat simpler than in fluorescent ballasts, due to the widespread use of single-lamp circuits.

(1) **Mercury Vapor Lamps** – For mercury vapor lamps, ballast circuits may be of the reactor, autotransformer, or regulator types. Regulator ballasts contain circuitry which is designed to operate the lamp at a relatively constant wattage even though nominal input line voltage may vary. As a rule of thumb, HID ballast losses range between 10% and 20% of lamp wattage. However, since each ballast circuit is somewhat different, catalog ratings should be used for more precise information.

(2) **Metal Halide Lamps** – Most MH lamps operate on a special ballast designed for metal halide lamps. The 1000 W type may be operated on a mercury vapor lamp reactor ballast if the ambient is over 50° F. The metal halide ballast (Fig. 4-8) is similar in circuitry to the mercury vapor lamp CWA ballast, but with modifications to provide the higher starting voltage required, and wave-shape characteristics to assure reignition of the arc each half-cycle.

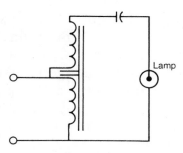

Figure **4-8**   Electric circuit diagram for metal halide ballast

(3) **High Pressure Sodium Lamps** – Since no starting electrode is incorporated in HPS lamps, the ballast must supply a high voltage pulse of 2500 to 4000 V at least once per cycle for wattages other than 1000 W. The 1000 W lamps require 4000 to 6000 V. The element that does this is called a starter or an ignitor. At present, four general types of ballasts are available to the HPS lamp users. Each has its advantages and disadvantages compared with the other in terms of lamp performance, cost, and energy consumption.

a)   *Reactor or Lag Ballast* – This can be made to have good lamp voltage regulation for changes in lamp voltages, but has poor regulation for changes in line voltage. It has a relatively high starting current, which can produce a desirably faster lamp warm-up. It is relatively inexpensive, having low power losses and is small in size.

b)   *Lead Ballast* – This has fairly good regulation for line voltage variations and also for lamp voltage variations.

c)   *Magnetic-regulated Ballast* – This is essentially a voltage regulating isolation transformer with its primary and secondary windings mounted on the same core, and containing a third capacitive winding, which adjusts magnetic flux with changes in either primary or secondary voltage. It provides the best wattage regulation with change of either input voltage or lamp voltage. It has a low line starting current and a high power factor. It is, however, the most costly and has the greatest wattage loss. Fig. 4-9 shows these three types of ballast circuits.

d) *Electronic Ballast* – The major problem with existing HPS lamp ballasts is their inability to operate the lamps at rated power. Electronically controlled ballasts have been designed and built with a solid-state control circuit and a reactor. The use of a solid-state switching device permits the control winding to be shorted in a "phase-controlled" manner, thus providing a smooth and continuous variation in the average inductance of the ballast. The solid-state control circuit monitors lamp and line operating conditions and then establishes the proper value of ballast required to operate the lamp at its rated power. Figure 4-10 shows an electronically controlled HPS ballast. The electronic ballast delivers efficiencies of well over 15% above that of the constant wattage type core coil-coil ballast. Further, it delivers additional energy savings of 13% by providing constant nominal lamp-rated wattage throughout the lamp life. The electronic ballasts have power factors around 99%, and produce line harmonics of approximately 5%.

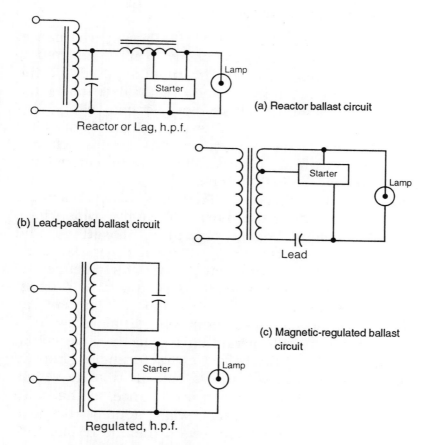

Reactor or Lag, h.p.f.

(a) Reactor ballast circuit

(b) Lead-peaked ballast circuit

Lead

(c) Magnetic-regulated ballast circuit

Regulated, h.p.f.

**Figure 4-9** Electric circuit diagram for HPS lamp ballast

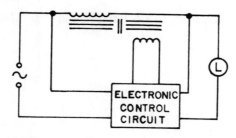

**Figure 4-10** Electronically controlled HPS ballast circuit

## Lamp Performance Factors

Ballast characteristics affect HID lamp performance. Ballast design determines the ability to start the lamp at low temperatures, controls the time required for the lamp to reach full output, and greatly determines the tolerance of a lamp to voltage dips. Serious voltage dips or any power interruption will extinguish the lamp, after which the lamp must cool for several minutes before it can restart. Figure 4-11 shows the warm-up characteristics of three types of HID lamps.

Because the voltage of HPS lamps is dependent on lamp wattage, unlike the other HID lamps, which have relatively constant voltage regardless of wattage, and since lamp voltage rises during life, lamp manufacturers have established trapezoid diagrams which define the lamp voltage and wattage limits. Figure 4-12 is a trapezoid diagram for a 400 W HPS lamp. For present-day ballasts operating at rated input voltage, all lamps of a particular design will operate on a smooth curve within the trapezoid called the ballast characteristic. A properly designed ballast will generate a smooth wattage versus voltage curve with a haystack appearance so that with increasing voltage the wattage will gradually rise to a peak, then start to decrease toward the end of life. HPS lamps on electronic ballasts can be made to traverse the trapezoid by way of a straight line. If the light output of such a combination is constant, input watts can be reduced approximately 20% over the life of the lamp.

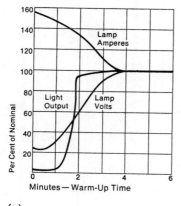

(a)

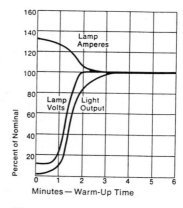

(b)

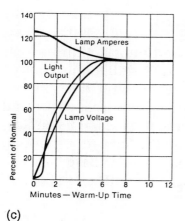

(c)

Figure **4-11** Warm-up characteristics of three types of HID lamps: (a) Mercury lamps; (b) Metal halide lamps; (c) High pressure sodium lamps.

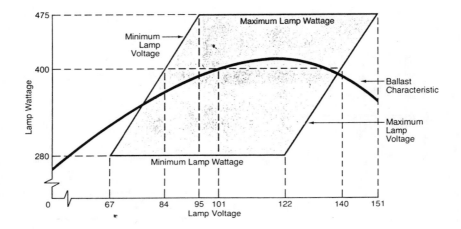

Figure **4-12** Trapezoid diagram for a 400 W high pressure sodium lamp

## 4.3 Luminaires

### 4.3.1 Types of Industrial Luminaires

There are many types of industrial luminaires. Selection of specific types for an installation requires considerations of many factors: candlepower distribution, efficiency shielding and brightness control, mounting height, lumen maintenance characteristics, mechanical construction, environmental suitability for use in normal, hazardous, or special areas. In general, there are five types of luminaires, in accordance with CIE classification for interior applications:

(1) **Direct type** – Direct type units emit practically all (90 to 100%) of the light downward to the working area. Although such luminaires usually provide the most

efficient illumination on working surfaces, it is usually at the expense of other factors. For example, shadows may be excessive unless the units have relatively large luminous areas or are mounted closer together than suggested maximum spacing-to-mounting height ratios. But direct and reflected glare may be disturbing because of the higher luminance difference between the bright source and the darker surround.

Direct industrial lighting equipment is usually classified according to the distribution of the downward component from "highly concentrating" to "widespread". This classification of luminaires is expressed in terms of suggested spacing-to-mounting height ratios, which are shown in Table 4-2. The widespread category includes high-intensity discharge (HID) luminaires that have optical assemblies consisting of a refractor/reflector design that can provide lamp concealment and reduce luminance sufficiently to permit a lower mounting height than would be acceptable for conventional HID luminaires. The distribution of low-bay units tends to improve vertical illumination (because of their wide-angle component) and to permit spacing as much as two or more times their mounting height above the work plane.

Table **4-2** Classification of Luminaire Direct Component Expressed in Terms of Space Criteria

| Spacing-to Mounting-<br>Height Ratio<br>(Above Work-Plane) | Luminaire<br>Classification |
|---|---|
| Up to 0.5 | Highly Concentrating |
| 0.5 to 0.7 | Concentrating |
| 0.7 to 1.0 | Medium Spread |
| 1.0 to 1.5 | Spread |
| Over 1.5 | Wide Spread |

Prismatic or mirrored glass or specular aluminum reflectors produce the more concentrating distributions. These are useful when luminaires for general lighting must be mounted at a height equal to or greater than the width of the room, or where high machinery or processing equipment necessitates directional control for efficient illumination between the equipment. They are also useful for supplementary illumination. Spread types are comprised of porcelain-enameled reflectors, other white reflecting surfaces, diffuse aluminum, mirrored or prismatic glass or plastic, and similar materials. Spread distributions are advantageous in low-bay areas or where there are many vertical or near-vertical seeing tasks.

Generally speaking, concentrating and medium spread distributions are best suited to high-bay areas. Wherever there is a need for higher-than-average general illumination for an inspection or special work area, highly concentrating luminaires should be installed above cranes at mounting heights where the basic high-bay lighting system is located. For large areas, low-luminance luminaires are preferred to provide low-reflected luminance. Such luminaires may consist of a diffusing panel on a standard type of fluorescent reflector, an indirect lighting hood, or a large luminous area. In a very dusty or corrosive area, luminaires with gasketed glass or plastic covers are recommended.

Area lighting extending from wall to wall is another form of direct lighting in which light from sources in a large cavity of high reflectance is directed downward through cellular louvers or translucent or refracting glass or plastic. When these materials conceal the lamps completely, the illumination characteristics are similar to those of indirect lighting system. Cellular louvers used as the shielding medium may present a reflected glare problem. This should be minimized in the design. Figure 4-13 shows a typical 400 W high-pressure-sodium luminaire that is popular in general factory illumination.

Figure **4-13** A typical 400 W HPS luminaire

(2) **Semi-direct Type** – These units emit 60 to 90% of their light downward. Utilization of light from these luminaires depends greatly on ceiling reflectance. Light-colored ceilings usually result in improved uti-

lization and visual comfort. The increased ceiling illumination from the semi-direct distribution reduces the luminance difference between ceiling and luminaire, increases diffusion, and softens shadows. Appropriately designed reflectors or refractors will reduce luminaire luminance and provide additional comfort. Most fluorescent and some HID and incandescent luminaires may be equipped with louvers to further increase shielding and reduce direct glare.

(3) **General Diffuse or Direct-Indirect Type** – In these luminaires, the downward and upward components are approximately equal: 40 to 60% of the total luminaire output. General diffuse type luminaires emit light about equally in all directions; direct-indirect luminaires emit very little light at angles near the horizontal, which is preferred because of their lower luminance in the direct glare zone. Luminaires with such a distribution are widely used in offices and laboratories, and their use in clean manufacturing areas is increasing.

(4) **Semi-indirect Type** – This type of luminaire emits most of the light (60 to 90%) upward. The major portion of the light reaching the horizontal work plane must be reflected from the ceiling and upper walls; therefore, it is necessary that these surfaces have high reflectance. The need for high reflectance and good maintenance limits the use of industrial semi-direct systems to areas where it is necessary to minimize reflected glare from specular work surfaces.

(5) **Indirect Type** – Indirect luminaires emitting from 90 to 100% of their light upward are seldom used in industry. These units have the lowest utilization and are more difficult to maintain.

Figure 4-14 shows luminaires for general lighting as classified by the CIE in accordance with the percentage of total luminaire output emitted above and below horizontal.

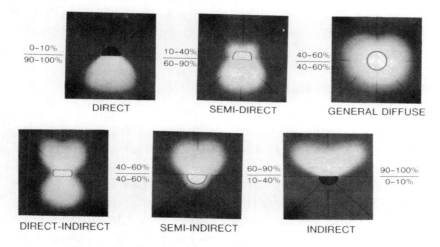

Figure **4-14** General lighting luminaire classifications

## 4.3.2 Supplementary Luminaire Types

Supplementary lighting units can be divided into five major types according to candlepower distribution and luminance:

(1) *Type S-I* – directional: includes all concentrating units, such as a reflector spot lamp or units employing concentrating reflectors or lenses.

(2) *Type S-II* – spread, high luminance: includes small area sources, such as incandescent or high-intensity-discharge. An open-bottom, deep-bowl diffusing reflector with a high-intensity-discharge lamp is an example.

(3) *Type S-III* – spread, moderate-luminance: includes all fluorescent units having a variation in luminance greater than 2:1.

(4) *Type S-IV* – uniform-luminance: includes all units having less than 2:1 variation of luminance. Usually, this luminance is less than 6800 candelas per square meter (2000 footlamberts). An example of this type is an arrangement of lamps behind a diffusing panel.

(5) *Type S-V* – uniform luminance with pattern: a luminaire similar to type S-IV, except that a pattern of stripes or lines is superimposed.

### 4.3.3 Specular Reflectors

In recent years, specular reflectors have been promoted as a potential source of energy savings for fluorescent lighting systems. A specular reflector is a luminaire component that has a highly polished surface. Applications of specular reflectors can increase luminaire efficiency, thus reducing the number of lamps, ballasts and/or luminaires that would be required in the system.

Specular reflectors can be used in new fluorescent luminaires or installed in existing luminaires as a retrofit strategy. In general, specular reflectors are made of one or more of three material types: anodized aluminum, enhanced anodized aluminum, and silver film that is applied to a metal subtrate. The long-term performance of the retrofitted luminaire is affected by its material's characteristics. Any degradation of material during its life will affect the luminaire's ongoing performance. Figure 4-15 shows the position of a specular reflector within a luminaire. The existing lamps are replaced with new lamps, and the luminaire and lens are cleaned. When these steps are taken, the connected power to the luminaire is approximately half of that prior to the retrofit, but the average illuminance may be greater than half by a substantial margin.

## Cross Section of Luminaire

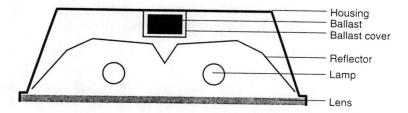

Figure **4-15** The position of a specular reflector within a luminaire

## Bibliography

Chen, Kao, Industrial Power Distribution & Illuminating Systems, Marcel Dekker, Inc., New York, N.Y., 1990.

Knisley, Joseph R., Updating Light Sources for New and Existing Facilities, EC&M, November, 1990, pp. 49-58.

Verderber, R.R., Morse, O., & Rubinstein, F.M., Performance of Electronic Ballast and Controls with 34 & 40W F40 Fluorescent Lamps, Proc. IEEE-IAS Conference, San Diego, CA, Oct. 1989.

NLPIP Specifier Reports on Electronic Ballasts, Rensselaer Lighting Research Center, Troy, N.Y., Dec. 1991.

NLPIP Specifier Reports on Specular Reflectors, Rensselaer Lighting Research Center, Troy, N.Y., July 1992.

NLPIP Specifier Reports on Compact Fluorescent Lamp Products, Rensselaer Lighting Research Center, Troy, N.Y., April 1993.

# Chapter 5

# Lighting Controls and Daylighting

## 5.1 Introduction

Lighting controls have become increasingly sophisticated in recent years, mainly due to the dire need for energy savings and advancements in the solid-state control devices. As the cost of energy has continued to rise, increasing effort has gone into minimizing the energy consumption of lighting installations. This effort has evolved along three major directions: the development of new energy-efficient lighting equipment, the utilization of improved lighting design practice, and finally, improvements in lighting control systems. The lighting industry has been introducing more efficient lighting components and systems, which cost more to install but result in a lower total cost (operating plus initial). End users begin to base their decisions on the ROI or the life-cycle cost. These decisions should lead to much greater application of cost-effective lighting control systems.

In the last decade, many new lighting control hardwares have appeared on the market. In this chapter we review all important types of lighting controls for industrial illuminating systems and offer some guidelines for selection to suit individual needs and achieve optimum energy savings.

## 5.2 Types of Controls

All lighting controls, whether a simple switch or a sophisticated programmable controller, can normally be classified in two basic categories: (1) on-off controls and (2) level controls. In its simplest form, lighting control can be accomplished manually by means of a switch located on a wall, in a luminaire or a panel box. Even though manual on-off switches for lighting control are used in commercial and industrial facilities, the current trend is toward greater use of lighting contactors. However, many circumstances require a varied level of illumination. Dimming devices are the most used means of providing the level controls.

Lighting controls can also be grouped into two general categories: centralized controls and local controls. The main difference between them falls into the realm of function of the area considered. Centralized controls are used in buildings where it is desirable to control large areas of the building on the same schedule. An example of centralized controls is a microprocessor that turns all lights on and off on a preprogrammed schedule. Localized controls are designed to affect only specific areas. Examples of localized controls are are personnel detector or a photocell controlling each office within a suite of offices.

Since a centralized system can be utilized to activate local controls, a time clock (centralized control) can be used to energize the building's entire lighting system on a time-of-day schedule, while a personnel detector (local control) located in a specific office overrides the centralized control to turn lighting in the specific offices on and off as demanded by occupancy. Figure 5-1 shows a matrix of the functions required for lighting control systems, with the type of control designed to perform that function or functions. No attempt is made to differentiate the actual technology utilized by manufacturers.

| Type of Control | Total Building Control | Localized Control | Programmable | On/Off | Manual Override | Telephone Control | Time of Day | Holiday/Weekend Schedule | Dimming | Co-ordination with Daylight | Remote Control | Co-ordination with HVAC | Over-the-wire Control | Low Voltage |
|---|---|---|---|---|---|---|---|---|---|---|---|---|---|---|
| **Centralized Controls:** | | | | | | | Functions* | | | | | | | |
| Computer operated | x | x | x | x | x | x | x | x | x | x | x | x | x | x |
| Over-the-wire controller clock | x | x | x | x | x | | x | x | | | x | | x | x |
| Programmable control | x | x | x | x | x | | x | x | | | x | x | | x |
| **Localized Controls:** | | | | | | | | | | | | | | |
| Daylight control (dimmable)¹ | | x | | | x | | | | | x | x | | | x |
| Daylight control (non-dimmable) | | x | | | x | | | | | x | x | | | x |
| In-fixture daylighting control | | x | | | x | | | | | x | x | x | | |
| Solid state dimming (no dimming ballast) | | x | x | | x | | | | x | x | | | x | |
| People sensors | | x | | x | x | | | | | | | | | x |
| Ambient light co-ordinators² | x | x | x | | | | | | | x | x | | x | |
| Timer control | x | x | | x | x | | x | | | | | | x | |

¹Requires dimming ballast in each luminaire
²Compensates for light loss factors
*Many functions can be added; check with manufacturer

Figure **5-1** Lighting control matrix

## 5.3 ON-OFF Controls

### 5.3.1 Wall Switches

A.C. snap switches can be used to permit selective use of lighting and also allow for different levels of lighting in a space. A variety of devices can be used, such as 3-way switches, to provide a path of lighting along a hallway. An illuminated switch is lighted when in the OFF position so it's easy to spot in a darkened room. A pilot light switch is lighted when in the ON position to serve as a visual reminder that a remote lighting load is on. A press switch allows no hands operation so even those carrying objects can cooperate in reducing lighting energy use.

Key activated AC switch can be operated only by a standard key, which is provided with the device, thus preventing unauthorized use of a lighting circuit.

### 5.3.2 Lighting Contactors

(1) **Power Lighting Contactors** – Power contactors can be divided into three general categories: power contactors rated up to 1200 A, multipole contactors with up to 12 poles, and single-pole contactors with low voltage control. More popular sizes range from 60 to 225 A. Since small contactors can be used for control of stations and an unlimited number of stations can be used for each power lighting contactor, the illuminating engineers can be liberal with the number of control stations used. A typical installation is the Twin Towers of World Trade Center in New York City. Four 225 A power lighting contactors are used for each floor. They are in turn controlled by two master controllers. Being mechanically held, they can be controlled by:

    a) A manually operated three-position switch with a center-off position
    b) Auxiliary relays as a function of photoelectric cells
    c) a time switch with a single-pole, double-throw contact
    d) Control relays in an energy management system (e.g., programmable controllers)

(2) **Multipole Lighting Contactors** – Multipole contactors are available up to 12 poles with contacts usual-

ly limited to a 20 A rating. The latest multipole contactors can be provided with optical solid-state control modules that provide two-wire control, three-wire
control, or stop/start control. A typical six-pole lighting contactor with a solid-state two-wire control module is shown in Figure 5-2.

Figure **5-2** Six pole 20 A lighting contactor
(Courtesy of Automatic Switch Company)

Today's multipole lighting contactors are of shal-
low construction, permitting them to be mounted in
a 4-in. stud-construction wall. Magnetically held 20
A multipole lighting contactors are also available.
They are often furnished, assembled and prewired, in
enclosures with electronic transceivers for connection
to programmable controllers that utilize card readers,
telephone interfaces, card printers, and other exter-
nal components as needed.

### 5.3.3 Low-Voltage Relays

Low-voltage relays are usually single-pole and used for individual branch-circuit or luminaire control. These relay contacts are normally rated for a 20 A tungsten filament load at 125 V ac and are mechanically latching, requiring only a momentary 24 V rectified ac switch circuit pulse to either open or close the local contacts. A step-down transformer is required to provide low-voltage power for relay actuation. With a 40 VA rating, one transformer can supply power to up to 15 relays with No. 14 AWG wiring.

The latching relay has three leads to provide the circuit path through the solenoid for latching or unlatching the line-voltage contacts. A second model has an internally energized pilot contact for a pilot light indication and so it requires four leads. A third model has an isolated internally energized pilot contact and thus has five leads. The five-lead relay with isolation of the pilot circuit is for special applications, such as wiring to a computer or a separate power source for the pilot light. In Figure 5-3 the solenoid coil is center tapped because the magnetic field strength required for unlatching is less than the magnetic field strength needed for latching.

### 5.3.4 Timing Controls

A broad variety of time switches are available for controlling lighting loads. They may be used for direct on-off control of lights or for control of lighting contactors. Twenty-four-hour dial timers are available with one or more sets of on/off trippers. A day-omitting device is optional for applications in which lighting is not used on weekends. Astronomic dial timers turn lights on at sunset and off at a prescribed time. Seven-day-calender dial timers are used when the on/off program is set on one

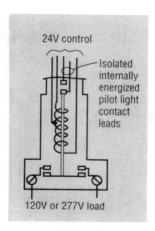

24V control

Isolated
internally
energized
pilot light
contact
leads

120V or 277V load

Figure **5-3** Low-voltage latching relay and wiring

dial. Operation can be omitted on selected days. Tripper will require periodic resetting to confirm to seasonal changes.

Program dial timers are used for multiple daily operations or for short-duration needs, such as turning lights on and off for cleaning crews and security guards. A total of 48 on/off operations can be programmed daily. They are available in multidial versions and with day-omitting capability. Reset timers incorporate a manual control and a timer. They are typically used by security guards on patrol; the guard can turn on selected lights when entering the area, and the timers turn off the lights after a prescribed time has elapsed.

Digital timers are used where multiple circuits, accuracy, and more sophisticated programming are required. They can be scheduled a year in advance for holidays and so on. However, they have a limited load-carrying capacity, typically 3 A at 24 V ac per circuit. This requires interfacing with a lighting contactor or relay and a step-down transformer. Figure 5-4 shows the front-panel illustration of digital timers for controlling eight lighting circuits.

## Programming Features

- Astronomic data can be conveniently copied into other channels without repeating settings.

- Only need to enter latitude to set astronomic feature.

- Flashing LED's prompt the user through correct sequence of settings

- Instant entry at the touch of a key

- Positive touch keys

- Audible signals confirm each setting and notify errors

- LED's visible from sides and distance

- Only one entry need be made for the same event on several different days

- Easy to go back and change a part of any entry

- All loads remain functioning during re-scheduling – then Z400 gives instant look back to execute the new program

- Automatically calculates day after entering date and year

Figure **5-4** Front panel of a digital timer with self-prompting LED features (Courtesy of Tork Co.)

Phototimers combine a remotely mounted photo-control and a seven-day dial timer. It controls three circuits, each with its own program. One circuit provides on and off operation by photocontrol; in the second, photocontrol turns the circuit on and a time switch turns it off; and in the third, on and off control are both provided by a time switch. Built-in manual bypasses maintain any circuit in on and off operation for prolonged periods without affecting the other circuits or the master program.

### 5.3.5 Sensors

Sensors for lighting controls can be divided into two categories: photoelectric sensing and presence detectors.

(1) **Photoelectric Sensors** – When used for control of outdoor lighting, photoelectric sensors are normally set to switch lighting on at dusk and off at dawn. Adjustments can be made to change response to higher or lower light levels. Built-in time delay helps to eliminate nuisance switching in response to sources other than natural ambient light.

Ambient lighting control systems using photocell sensors can serve switching and continuous dimming functions. In response to the photocontrol unit switching system will turn on lights in a specific pattern to provide typically 1/3, 1/2, or 2/3 of full light output in an area. A continuously dimmable system will adjust the output of the electric lighting system in response to the amount of daylighting striking the control photosensor.

(2) **Personnel Sensors**

A variety of personnel sensors have recently been developed. Personnel sensors are complete control systems that sense when a space or room is occupied and automatically turn the lights on for a preset period. If no further occupancy is sensed during this time interval, the lights are then turned off automatically. There are many types of personnel sensors; they differ only in their method of sensing occupancy. Some of these methods include passive and active infrared, ultrasonic, and acoustic sensing.

a) *Passive Infrared Sensors* – They detect and respond to changes in radiated heat within a room caused by the presence and movement of a human body. They consist of two components, the sensor and the control unit. The sensor, which contains the passive sensing-element optics system and the electronic logic circuitry, is designed to be mounted in the ceiling panel. The control unit, which contains the low-voltage power supply for the system and the load relay used to switch the lighting load, is typically mounted above the ceiling. The sensor includes a timer circuit that keeps the lights on as long as changes in infrared energy are detected. If no changes are detected for a certain period, the lights are turned off automatically.

The most common passive infrared sensors have a coverage area of approximately 200 ft². If a larger area needs to be covered, multiple sensors can be connected to a single control unit, as shown in Figure 5-5, which indicates placement of three 200 ft² sensors for coverage of a 20 ft by 20 ft L-shaped office.

There are design considerations in the use of personnel sensors. The sensor must be placed so as to cover areas of the room where occupants are expected to be. Care must be taken to ensure that the sensors cover all potential occupant locations. The sensor should not be placed to close to the entry; otherwise, people walking in an outside corridor past an open door will activate the lights. Passive infrared sensors should also be placed where they will not sense any nonhuman heat sources, such as an HVAC register or baseboard heater.

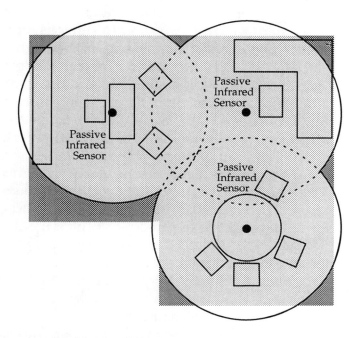

Figure **5-5** Placement of sensors for coverage of 400 ft. office

b) *Ultrasonic Sensors* – An ultrasonic sensor creates an ultrasonic field in the room being monitored. When a person enters the room, the field is disturbed; the sensor detects the disturbance and activates the room lights. A typical ultrasonic sensor unit containing all the sensing and local control equipment is mounted on the ceiling, and can provide up to 900 ft2 of coverage. A single unit often contains multiple sensors.

Each type of ultrasonic sensor has a coverage pattern that depends on the number of sensors mounted in the unit. The pattern may include two different coverage areas for the same unit: an area within which large body motions, such as walking, are detected, and a smaller coverage area, within which small body motions, such as the movement of an arm, are detected. Careful placement of the ultrasonic sensors ensures proper operation of the system. Sensors must be placed such that they do not detect motion outside the room being controlled. Another consideration in the application of ultrasonic sensors is the acoustics of the room. Room acoustics can affect the coverage pattern and therefore the appropriate number and placement of sensors.

Ultrasonic sensors are available in many different styles, each providing specific coverage patterns. There are also ultrasonic sensors designed for wall mounting as a direct replacement for a doorway light switch. Personnel sensors are ideal for controlling lights in any space with random and intermittent occupancy patterns. Some spaces that meet this criterion include enclosed offices, rest rooms, storage rooms.

c) *Acoustic Sensors* – They respond to sounds created as a result of a person's movement in an area. The unit is useful where direct line of sight to a sensor can not be achieved, such as an irregularly shaped room or corridor, a stairway, or a storage room.

d) *Active Infrared Sensors* – They emit invisible infrared light beams in a specific pattern and a receiver responds to changes in the light beam patterns caused by a person's movement in the field of view.

## 5.3.6 Programmable Control System

Control systems are now available that employ microprocessor logic to replace hard wiring with soft wiring. Coded comments can be multiplexed to control points over a pair of low-voltage wires. Control points have receiver/switches, the latter complement generally a low-voltage relay or lighting contactor. Logic functions can be programmed into a control device to turn lights on and off over a 24-h or 1-week period. Overrides are available, and some systems can be accessed with Touchtone telephones.

Such Systems, which control lighting in both time and space, save considerable energy compared with past control practices. The heart of such systems is the programmable controller, which holds in memory a series of on-off instructions to the lighting circuits throughout a building. They can provide minute-by-minute control of an entire lighting scheme according to a user-determined schedule, with pulse initiation of the control signal generated from its internal clock.

### 5.3.7 Control Mediums

A number of methods are used to provide area-wide control in a building:

(1) **Centralized Programmable Control** – Multiplexed signals from the programmable controller can be sent through a building via twisted low-capacitance wire. Signals can also be sent via existing power wiring to the receiver control modules. The codes from the controller are transmitted to the transceivers, where they may be combined with other signals (e.g., photo-control relay output). In turn, signals are provided to low-voltage relays or lighting contactors, which require only a momentary electrical pulse to operate its latching or mechanical mechanism to the on or off position.

Control of the lighting pattern is not limited to the programmable controller keyboard. In addition, a manually operated, momentary contact switch connected to the transceiver can provide signals to operate not only those relays, or contactors connected to it, but also relays or contactors located anywhere throughout the building via the data line back to the programmable controller.

(2) **Power-Line Carrier System** – The power-line carrier system contains electronic transmitters and receivers that use the building power wiring as a communications pathway. The transmitter accepts a control signal input, converts the control signal into digital form, and injects it onto the power wiring system. The low-voltage digital signal is transmitted at a frequency anywhere from 25 to 250 kHz, depending on the specific transmitter design. It is important to note that host signals cannot pass through transformers; bypass devices are usually required at each transformer.

(3) **Radio-controlled Ballast System** – One interesting use of power-line carrier technology is a high-intensity-discharge (HID) dimming ballast that has a receiver built into it. Control signals carry dimming information to all ballasts equipped with integral receivers. This system can be connected to a photosensor to provide for either daylighting compensation or lumen maintenance control. A power-line carrier system can be economically used in large facilities with many control points, such as factories and warehouses where control cabling would be lengthy and expensive. The technology is also suitable for existing buildings, where the cost of installing new lighting control wiring can be prohibitive.

## 5.4 LEVEL Controls

### 5.4.1 Dimmers

Many circumstances require a varied level of illumination. Dimmers are the most used means of providing lighting level control. The original dimmers were of the resistance types. In the last decade, solid-state dimmers have taken over 90% of the market. By means of an electronic switch, the electronic dimmers turn off the current to the load for a portion of the cycle, thus delivering less power to the load. Electronic dimmers are now available for incandescent, fluorescent, and HID lighting.

### Types of Dimmers

(1) **Conventional Dimmers** – The modern SCR (silicon-controlled rectifier) dimmer operates on the principle of switching on the current a proportional distance through each half-cycle. An SCR is nothing more than a very fast switch. Dimmers operate simply by delaying the turning on of these switches by an amount of time inversely proportional to the incoming control voltage. In the simplest of examples, a dimmer set at 50% will delay halfway through each half-cycle and then turn on.

(2) **Digital Dimmers** – It uses the same very fast switch (SCR), but instead of using a voltage to charge a capacitor and turn on the switch, the digital dimmer counts a number of steps through each half-cycle, then turn on the switch.

Another term that is often used when referring to new technology dimmer is multiplexing. Multiplexing refers to the use of a single cable to carry data to a group of dimmers. Multiplexing is a concept that can be used on both analog and digital dimmers.

(3) **Intelligent Dimmers** – They communicate with the controller in the same way as does the digital dimmer except that they have the intelligence to recognize data other than the setting of the controller. The scope of what these data could include is as broad as the scope of what the intelligent controller can produce.

## 5.4.2 Technology Of Patching

As dimmer per light system becomes more and more clearly the wave of the future, moving the job of patching out of the controller and into the dimmer makes more and more sense. If in this data stream the intelligent dimmer is being informed that it is patched to channel 5 and that whenever it sees data for channel 5 it should use them, the number of dimmers that can be patched to any control channel will be limitless.

One problem when lights are patched together in a single control channel is that the light produced when that channel is active may not be uniform across an area. If a proportional level is part of the data that the intelligent dimmer stores and uses, all the dimmers patched to one channel can be balanced to make the lighting uniform. It is possible for an intelligent dimmer to communicate information gathered back to the controller so that the controller can make adjustments and/or communicate this information to the operator. For years, conventional analog dimmers have used feedback information to adjust the operation of dimmers – line regulation, load regulation, and current limiting being most common.

The intelligent dimmer will be able to do as many of these functions as engineers deem desirable. The intelligent dimmer will run a program like any other computer, and this program can be changed. The scope and capabilities offered by it are going to be important parts of control technology in the future.

### 5.4.3 Computer/Microprocessor

Most sophisticated lighting control now consists of a microcomputer, an oscillator as its controller and a receiver switch as decoder. The microcomputer is programmed for various lighting patterns and addresses of the luminaires. The address and condition codes generated are modulated by the oscillator and superimposed on the building electrical system line frequency. At the receiver/switch, a decoder takes the message off the line, and if the address code corresponds to the one given that switch, the condition codes are executed and turn the luminaire fully on, halfway on, or off. One such system contains all programming, logic circuitry, and hardware. Discretionary control and override functions can also be incorporated. Printer can be connected to the microcomputer to provide hard-copy records of all control activities. This feature gives information to the system operator about the frequency and time of any local override activities. Video display terminals and recent software have made the system much easier to use, which promotes prompt revision of control schedules. Transceivers are available in 16- and 32-output versions. The control relays are capable of switching a 20 A inductive load. Figure 5-6 shows a typical scheme for such a control system as just described.

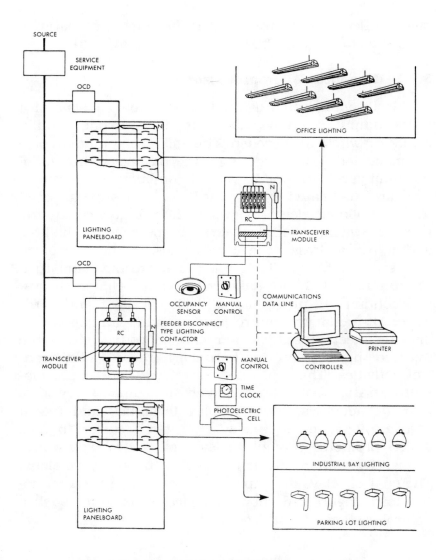

Figure **5-6** Programmable lighting control scheme
(Courtesy of Automatic Switch Company)

## 5.4.4 Manual Incandescent Dimming Devices

(1) **Wallbox-Type Dimmers** – They are rated up to 2000 W of lighting load to be manually controlled from a single electrical device box. Some models allow dimming from two or more locations, which previously required a remotely controlled dimmer module. Wallbox dimmers are also available with integral electronic switching to allow remote ON/OFF control from multiple locations, using controls that aesthetically match the dimmers.

(2) **Microprocessor-based Wallbox Dimmers** – They combine dimming control, preset memory of a desired lighting level, and time clock operation. A typical unit containing four dimmers, each with four scenes of possible settings, fit a standard multigang device box.

(3) **Console-sized, Microprocessor-based Systems** – They can provide memory for hundreds of scenes and lighting zones, and the automatically produced dimmer settings can change at very slow fade times.

## 5.5 Bases For Selecting Lighting Controls

### 5.5.1 Guidelines for Minimum Number of Lighting Controls

Table 5-1 represents an attempt to establish some guidelines for a minimum number of lighting controls to be installed with a reference to the area size and unit power density of a lighting system. The lighting controls used are either on-off devices or dimming controls.

Table **5-1** Recommended Minimum No. of Lighting Controls*

|         | UPD (W/ft) | | |
| A (ft)  | 1.5 | 1.5-3 | 3 |
|---------|-----|-------|---|
| 125     | 1   | 2     | 2  |
| 125-250 | 1   | 2     | 2+ |
| 251-500 | 1   | 2+    | 4  |
| 501-1000| 2   | 3+    | 5+ |
| 1001-2000| 3  | 4+    | 6  |
| 2000    | 3+  | 5+    | 6+ |

\* When multiple controls are used, it is generally installed
to permit reducing the general lighting in the space by at
least one-half in either a uniform pattern or by zones, as
most appropriate.

## 5.5.2 Medium-Sized Offices

An assessment of the cost-effectiveness of photoelectric control equipment was made for three medium-sized offices.

(1) A fully automatic mixed control system would be unlikely to be cost-effective for single offices.
(2) A partially automated on-off system would be cost effective in new buildings at present energy costs.
(3) A partially automated mixed system for new buildings would probably be cost-effective only if designed to control luminaires in several offices.

The results of this study are summarized in Table 5-2. From a close examination of Table 5-2 it would seem sensible to include a dimming line to each luminaire during its installation in new buildings. This would involve only a small extra capital cost, but would keep open the options of installing a dimming system later without the necessity for complete rewiring. In another study, when due considerations were given to daylight factor, it appeared that it is generally best to control only those luminaires nearest the windows.

Table **5-2** Cost-Effectiveness of Photoelectric Control Systems+

| Type of control | New Buildings | | Existing Buildings | |
|---|---|---|---|---|
| | Single office | Multi-office | Single office | Multi-office |
| Fully automated | | | | |
| 1 on-off + 1 dimming | 2 | 1 | 2 | 2 |
| Partially automated | | | | |
| 1 on-off | 1 | 1 | 3 | 3 |
| 1 on-off + 1 dimming | 2 | 1 | 2 | 3 |

+ 1)   Cost-effective within 15 years; 2) not cost-effective within 15 years;
3) depending on energy costs.

### 5.5.3 Factors Affecting Selection of Lighting Controls

In general, there do not appear to be any general rules or guidelines that conveniently lead one to select specific controls. The following factors will have a bearing on the selection of lighting controls:

(1) **Size of Facility** – A large facility may justify a building management system or programmable controllers that provide centralized lighting control. On the other hand, a small facility may obtain optimum savings by selecting a simple time switch control. From experience, large computers often cost around $500 per hour to operate, whereas a minicomputer costs no more than $10 per hour to operate.

(2) **Size of Individual Lighting Area** – If the current drawn by an individual lighting area exceeds 20 A a power contactor may be the choice. Small areas such as individual offices would be candidates for low-voltage relays. In either case, these devices would be controlled by timers, photoelectric sensors, and the like. In recent years, the trend has been to control smaller individual lighting areas.

(3) **Availability of Daylighting** – As discussed previously, energy savings from daylighting depends on many factors: climate conditions, building form and design, and the activities within the building. Only a portion of a building can be daylit; however, in most cases, 30% of the floor space is sufficiently close to the perimeter to be daylit. There is little documented research in this field. As a general guide, the average energy savings from daylighting for an entire building will be in the neighborhood of 15%.

(4) **Type of Usage in the Facility** – If the facility provides commercial rental of office space, consideration is often given to flexible controls, such as low-voltage relays. In the institutional facilities where lighting requirements are more fixed, other types of controls should be considered.

(5) **New Installations or Modification of Existing Facility** – From Table 5-2 it becomes evident that a more elaborate photoelectric control system or even a sophisticated lighting management system can be easily justified for a new building. However, for an existing building, an extensive lighting control system may not be cost-effective. In this case a control system that utilizes existing power wiring for signal transmission would be preferred.

### 5.5.4 Comparison of Lighting Control Systems

Table 5-3 shows a simple comparison of different systems according to how well each meets the needs of both the occupant and the building management. The relative performance rating of each system is often a judgement call. Individual manufacturers for a single system may vary sharply in system performance and cost. The objective here is to show a process, not pass final judgement or provide hard pricing guidelines. Some of the reasoning that went into the relative rating for each function shown in Table 5-3 is given below:

Table **5-3** Comparison of Various Lighting Controls

| | Occupant Sensitivity | Light Level Selection | Energy Savings Potential | Manage-ment Data | Inte-gration Capability | Space Adapta-bility | Cost |
|---|---|---|---|---|---|---|---|
| Standard Wall Switches (Individual offices) | Good | Fair | Fair | No | No | Poor | Medium |
| Contractor Control Via Building Automation System | Poor | Poor | Poor | Yes | Yes | Good | Low |
| Programmable Lighting Control Relay Based | Excellent | Fair | Good | Yes | Yes | Good | Medium |
| Programmable Lighting Control Dimming Based | Excellent | Excellent | Excellent | Yes | Yes | Good | High |
| Occupancy Sensor | Fair | Poor | Good | No | No | Poor | High |

(1) **Occupancy Sensitivity** – From an occupant's perspective, individual wall switches work just fine. Contactors on most building automation systems can be a real disadvantage, since they do not normally allow the occupant to override for after-hour usage. Programmable lighting control caters to the occupant. When he or she enters the building after hours, the person's particular working area can be lit in anticipation of his or her arrival with a single phone dialing. Similarly, when a person is staying late, his or her office and related work space can both be kept on with a phone dialing or switch override.

(2) **Occupant-Level Selection** – This is a function affected by both control system capacity and floor layout. Individual office layouts with manual switching of split-wired fixtures or several lighting sources within the space give most occupants the degree of control they need. The dimmable solid-state ballast approach provides an even better method for allowing the occupant to adjust the overhead lighting.

(3) **Energy Saving Potential** – The only surprise here is in the "good" rating for switches in individual offices and the "poor" rating for the same devices used to control a zone. In practice, what happens is that when more than one person is in a zone, the first in turns it on and the last out never looks back.

(4) **Management Data** – This reflects system monitoring analysis, and reporting capabilities. These require a communications capability not normally inherent in switches or occupancy sensors.

(5) **Space Adaptability** – The important point here is that devices physically linked to the occupant's walls or ceiling pose problems when it is time to rearrange a space.

(6) **Costs** – The solid-state ballast cost represents a combination of functions not presently available on the market. Perhaps the biggest surprise is that individual office switches are not cheap. They typically cost $0.50 per square foot. Occupancy sensors may reduce the installation labor, but the added hardware content still means a relatively high total cost. Both switches and sensors incur added cost for office rearrangements.

In conclusion, regardless of the type and/or size of the facility which illuminating engineers may deal with, there will always be a suitable type of control for them to choose from. The engineers must become knowledgeable in lighting controls and exercise their sound judgement in the selection of control schemes to achieve a high quality lighting system with optimum energy savings.

## 5.6 Daylighting

### 5.6.1 General Discussions

Contemporary lighting design practices require that illumination meeting a specified design criterion be provided whenever building is occupied. The sole use of daylighting does not meet such a criterion, because daylighting illumination levels change continually with time, season, and sky conditions. To exploit daylighting as a source of illumination, it is necessary to establish an interactive link between ambient lighting conditions and the electric lighting system. This can be achieved with a photoelectrically controlled lighting system that adjusts the output of the electric lighting system based on the amount of prevailing daylight. Lighting control hardware that links electric lighting to available daylighting generally falls into two categories:

(1) Continuously dimmable electric lighting systems controlled by photosensors, which continually adjust the electric lighting level in response to the amount of daylight striking the control photosensors. At the heart of the system is the controllable output ballast or dimming ballast.
(2) Photo-relay-based systems, which automatically switch off perimeter lighting when daylighting is sufficient to meet lighting needs.

The lighting control system shown in Figure 5-6 can be used to achieve the above described lighting level control scheme with the aid of photosensors located within the task zones. Photosensor information is used by the microcomputer to control zone lighting levels, which can be controlled from 100 to 0%. In zones where natural light is available, the system automatically takes advantage of this resource and incorporates it into the space. The electric lighting in daylight zones is used as fill-in lighting to provide even illumination across the space. The amount of usable daylight is dependent on the windows and skylight types and sizes, window and skylight treatments (drapes, blinds, glazing, etc.), building design, interior design, and furnishings.

### 5.6.2  Designing for Daylight

(1) **Introduction** – In developing a preliminary building design, architects must first establish interior visual criteria and basic lighting performance requirements for their particular project. Next, they will have to determine the parameters of the available daylight for the particular location and select the appropriate daylight data to be used as a basis for design purposes. They are now ready to calculate daylight contributions for various design schemes.

(2) **Establishing a Sky Condition for Design** – Sky conditions are extremely variable for almost all localities. From morning to evening, from season to season, and from day to day, the amount of light available from the sun and sky is constantly changing. Additionally, very little measured data is available. It is impossible to establish a single set of conditions that can be used as an absolute design base. Generally, the designer will want to deal with maximum and minimum sky conditions or with the condition which may be considered most prevalent.

(3) **Predicting Interior Daylighting** – Interior illumination is determined for each of several conditions and then added together for final results. For instance, illumination levels from windows and from skylights can be determined separately, and then added together for final results. Illumination from the sky and ground are determined separately and added for final results. More detailed discussions are given below:

a) *Combination Daylighting Sources* – Daylighting designs which combine side-wall fenestration in opposite walls, or designs which combine side-wall lighting and toplighting can be treated by superimposition. That is, calculate the workplane illumination from one source of daylight; calculate the illumination from the second source; and add the two results to get the total illumination on the workplane.

b) *Combination Daylight and Electric* – Similarly, building designs which combine daylighting and electric lighting can be treated by superimposition. The workplane illumination calculated for daylight can be added to that calculated for electric light.

c) *Reflected Ground Light* – Although the condition of the ground varies from side to side and, thus, a part of the designed environment, in these calculations light reflected from the ground is considered as another light source. The light from the ground and from the sky must be separately determined.

d) *Other Prediction Techniques* – The daylight prediction technique described in the above is based on the lumen method commonly used for calculating the illumination from electric lighting and should be familiar to illuminating engineers. However,

there are a number of other techniques for predicting interior daylighting including graphic,
mathematical, and simulation processes.

### 5.6.3 Practical Method of Utilizing Daylight

A practical method of utilizing daylighting to conserve
energy is shown in Figure 5-7. Indicated herein is a
method of wiring two luminaires with a two-lamp ballast
to achieve three-step control. As more daylighting is
available, more rows near the window are switched off.
They may appear to be a crude method, but it is an economical approach that requires no elaborate control
hardware. The wiring method responds to daylighting in
the following manner:

| Sequence | Illumination (%) |
| --- | --- |
| Power on to ballasts X and Y | 100 |
| Switch off power to either X or Y | 50 |
| Switch off power to both X and Y | 0 |

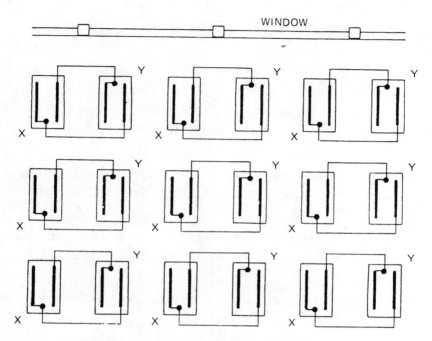

One two-lamp ballast controls one lamp in the luminaire and one
lamp in the adjacent luminaire. Illumination of each row can be adjusted in 3-steps in response to daylighting in the following manner:

| SEQUENCE | ILLUMINATION |
|---|---|
| Power on to ballasts X and Y | 100% |
| Switch-off power to either ballast X or Y | 50% |
| Switch-off power to both ballasts X and Y | 0% |

As more daylighting is available, more rows near the window are
switched off.

**Figure 5-7** Method of wiring for three-step control

## Bibliography

Alling, W.R., The Integration of Microcomputer and Controllable Output Ballast – A New Dimension in Lighting Control, IEEE Transactions on Industry Applications, vol. IA-20, no. 5, Sept./Oct. 1984.

Chen, Kao, and Castenschiold, Rene, Selecting Lighting Controls for Optimum Energy Savings, IEEE-IAS Annual Conference Proceedings, Oct. 1985.

Crisp, V.H.C., Preliminary Study of Automatic Daylighting Control of Artificial Lighting, Lighting Research and Technology, vol. 9, no. 1, 1977.

Pearlman, Gordon W., The Emergence and Future of Intelligent Dimmers, Lighting Design and Application, June 1982, pp. 22-23.

EC&M, Lighting Controls Help Reduce Energy Consumption, Nov. 1990, pp.61-64.

IES RP-5 Recommended Practice of Daylighting, Illuminating Engineering Society of North America, New York, N.Y., 1979

NLPIP Specifier Reports on Occupancy Sensors, Rensselaer Lighting Research Center, Troy, N.Y., Oct. 1992.

# Chapter 6

# Achieving and Evaluating Energy Effectiveness In Lighting Systems

## 6.1 Achieving Energy Effectiveness in A New System

### 6.1.1 General Discussions of Industrial Seeing Task

Industry encompasses seeing tasks, operating conditions, and economic considerations of a wide range. Visual tasks may be extremely small or very large; dark or light; opaque, transparent or translucent; on specular or diffuse surfaces; and may involve flat or contoured shapes. With each of the various task conditions, lighting must be suitable for adequate visibility in developing raw materials into finished products. The speed of operations may be such as to allow only minimum time for visual perception and therefore, lighting must be a compensating factor to increase the speed of vision.

The lighting system should be a part of an overall planned environment. The design of a lighting system and selection of equipment may be influenced by many economic and energy related factors. For related information, reference should be made to Chapters 3 and 4.

## 6.1.2  Quantity and Quality of Illumination

Illuminance recommendations for industrial tasks and areas are given in Tables 2-1 and 2-2. However, illuminance value for specific operations can be determined using illuminance categories of similar tasks and activities found in those tables, and the application of the appropriate weighting factors in Tables 2-3 and 2-4. In either case, the values given are considered to be target maintained illuminances. To insure that a given illuminance will be maintained, it is necessary to design a system to give initially more light than the target value.

The quality of illumination pertains to the distribution of illuminance in the visual environment. Glare, diffusion, direction, uniformity, color, luminance and luminance ratios all have a significant effect on visibility and the ability to see easily and quickly. For more detailed discussions on these subjects, reference should be made to Chapter 3.

Industrial installations of very poor quality are easily recognized as uncomfortable and are possibly hazardous. The cumulative effect of even slightly glaring conditions can result in material loss of seeing efficiency and undue fatigue.

### 6.1.3 General Design Considerations

The illuminating engineers who are to design an industrial system should consider the following as the first and all-important requirements:

(1) Determine the quantity and quality of illumination desirable for the manufacturing processes involved.
(2) Select lighting equipment that will provide these requirements by examining photometric characteristics and mechanical performance that will meet

installation, operating, and actual maintenance conditions.

(3) Select and arrange equipment so that it will be easy and practical to maintain.

(4) Apply control and daylighting techniques to the proposed system for promoting overall energy effectiveness.

(5) Balance all of the energy management considerations and economic factors including initial, operating, and maintenance costs vs. the quantity and quality requirements for optimum visual performance.

## 6.2  Achieving Energy Effectiveness in A Retrofit System

When specifying a retrofit system, the goal is to maximize energy efficiency in addition to meet the necessary lighting requirements for the occupants in the area. The system approach requires a thorough review of all aspects of the lighting system to be retrofitted. Each area of the facility should be studies to determine the task performed. An examination should be conducted of the existing lighting components that currently provide the illumination for these tasks. Once that survey has been completed and evaluated, appropriate retrofit systems can be specified for each area. Proper design based on the steps outlined in Chapter 2 should follow. To make sure that the proposed system satisfy the objective of energy effectiveness, the engineers must also not forget the application of the control and daylighting techniques to the system wherever feasible.

The following outlines the necessary steps to be carried out for this type of retrofit projects:

(1) **The Survey** – Accurate information is needed on what exists and how it is being used. Survey form

are useful and data can be collected on a hand-held dictating machine and transcribed to a spreadsheet or a database for analysis.

Each system must be noted as to its type, designated by the name of the representative fixture for each system, the type of lens or louver used, the method of mounting, the lamp type, the number of lamps per fixture, the type of ballast and number per fixture, the type control for each fixture subsystem, the tasks performed in the area, the type of ceiling. To determine whether or not certain fixtures can be tandem wired if multi-lamp electronic ballasts are to be used. The engineer should also find out if there are any special occupancy use requirements, the age of the occupants, whether computers or other video display terminals are used in the space. Levels of ambient illumination can be reduced if the primary work is on a computer screen or suitable task lighting may be applied.

(2) **Specify the Retrofit** – Once all the above data has been collected, the most efficient way to optimize the lighting system in each area should be determined. The proposed retrofit should minimize the energy input required to produce the appropriate illumination. In carrying out lighting retrofit project, some of key functional and technical performance factors as listed below should be carefully considered:

| *Functional performance* | *Technical performance* |
|---|---|
| • Task requirements | • Lamp lumen depreciation |
| • Illuminance values | • Efficacy or efficiency |
| • Color rendition | • Energy density |
| • Daylighting | • Life cycle cost |
| • Aesthetics | • Maintenance |
| • Controls | • Power factor |
| • Glare | • Harmonics |

It must be remembered that a) lighting levels will decrease or diminish with time, and b) a total fixture upgrade is a more effective approach. The use of occupancy sensors should be only considered for right application. Task lighting in the work area should also be considered as this may allow for a substantial reduction in ambient lighting. Another area to investigate is daylighting which can be used to supplement existing lighting systems and has become an effective means of saving energy.

(3) **Implementation** – When retrofit system design is complete, the next step is to focus on buying and installing the necessary components. Separately bidding the materials and the installation usually offers project cost savings. Special care should be taken to specify responsibility for disposal of PCB-containing ballasts and mercury-containing fluorescent and HID lamps. In addition to the materials and installation bids, engineers should also contact the local utility to find out the extent of its financial assistance to the lighting retrofit. Some utilities negotiate rebates. Each utility may have a slightly different program for rebate.

## 6.3  Evaluating Energy Effectiveness in A New or Retrofit System

### 6.3.1  Evaluating Methods

Energy effectiveness in an illuminating system carries a broader significance than energy efficiency of the various components which make up the system. As discussed in preceding sections, energy effectiveness can only be achieved by applying and implementing the following:

(1) Accurate determination of the seeing task requirements.
(2) Proper design approach to avoid any of the pitfalls (outlined in Chapter 3) to be introduced in the work environment.
(3) Optimum selection of the energy efficient components including suitable controls for the proposed system.
(4) Appropriate integration of the daylighting into the system wherever applicable.

Energy management and energy effective design have a tremendous impact on cost. The final decision as to which system to install depends heavily on the cost. These costs should include not only the initial cost of the installation, but also the operating and maintenance costs. With the rise in energy cost and inflation in all sectors of economy, an inexpensive system (low initial cost) could cost the owner many times more to operate and maintain.

(1) **Initial Cost** – Energy efficient design will result in an in crease of initial cost over the traditional approach of "make it cheap". Many factors are involved, so a decision as to which system to use should not be made on the basis of initial cost alone.

(2) **Maintenance Cost** – This is tied directly to the selection of lighting equipment, which affects initial cost. In general, the initial cost will increase for equipment with better maintenance characteristics. A good maintenance program will minimize light loss from dirt accumulation and surface deterioration, which would avoid increasing light to compensate for such losses.

(3) **Operating Cost** – This is tied into the amount of power consumed. System design in terms of light

source efficacy and overall system efficiency will determine the operating cost.

### 6.3.2 Cost Analysis

There are a number of methods of economic analysis by which the choices available may be reviewed. Types of analysis presently in use are the payback period, the internal rate of return, present value, savings investment ratio, and life-cycle costing. With the rapid increase in the cost of energy in recent times, an inflationary factor is critical to an analysis of operating costs. Two of the methods, namely, Payback Period and Life-Cycle Costing will be discussed more fully in the following:

(1) **Payback Period Method** – A major lamp manufacturer designed an energy cost management analyzer that contains all essential lamp information and fixture data, and provides parallel working spaces for the existing lighting system and the new energy effective lighting system, side by side. Each contains items such as annual energy cost per fixture, annual lamp replacement cost per fixture, annual cleaning cost per fixture, and so on. The total of these items is equal to the annual operating cost for the existing or the proposed system. From the difference between the two systems, one will be able to evaluate return on investment (ROI). Figure 6-1 shows such an analyzer.

There are many other forms that one can use to make a retrofitting study. However, the form prepared by IES and/or major lighting manufacturers is recommended to use for its thoroughness and completeness in essential information required for evaluation. Figure 6-2 shows such a form. Today many lamp and luminaire manufacturers offer computer services for evaluating lighting system alternatives.

# Your Present Lighting System

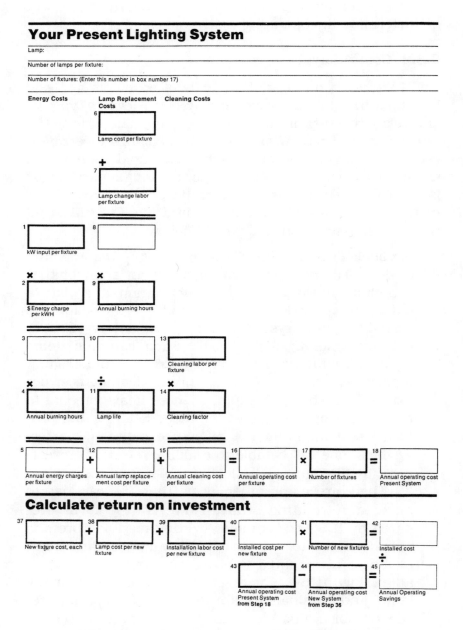

Figure **6-1** Lighting energy cost management analyzer

# Your New Energy Saving Lighting System

Lamp:

Number of lamps per fixture:

Number of fixtures: (Enter this number in box number 35)

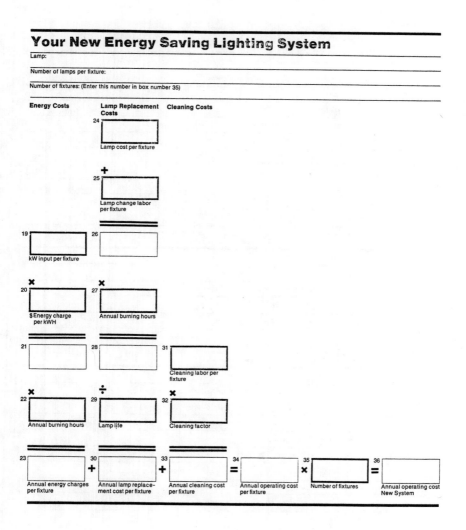

Figure **6-1** (continued)

| | | Lighting System Parameter | Base | II |
|---|---|---|---|---|
| **Basic Data** | 1. | Rated initial lamp lumens per luminaire | | |
| | 2. | Rated lamp life (hours) at _____ hours per start | | |
| | 3. | Group replacement interval (hours) | | |
| | 4. | Average watts per lamp | | |
| | 5. | Input watts per lumaire (including ballast losses) | | |
| | 6. | Coefficient of utilization | | |
| | 7. | Ballast factory (fluorescent) | | |
| | 8. | Lamp depreciation factor | | |
| | 9. | Dirt depreciation factor | | |
| | 10. | Effective maintained lumens per luminaire $(1 \cdot 6 \cdot 7 \cdot 8 \cdot 9)$ | | |
| | 10A. | Average footcandles on work surface $(10 \div ft^2/\text{luminaire})$ | | |
| | 11. | Relative number of luminaires needed for equal maintained footcandles (10 of base system $\div$ 10 of system compared) | | |
| **Initial Costs** | 12. | Net cost of one luminaire | | |
| | 13. | Wiring and distribution system cost per luminaire | | |
| | 14. | Installation labor cost per luminaire | | |
| | 15. | Net initial lamp cost per luminaire | | |
| | 16. | Total initial cost per luminaire $(12 + 13 + 14 + 15)$ | | |
| | 17. | Annual owning cost per luminaire (15% of $12 + 13 + 14$) | | |
| | 18. | Relative initial cost for equal maintained footcandles $(16 \cdot 11$ of system compared $\div$ 16 of base system) | | |
| **Operating Costs** | 19. | Burning hours per year | | |
| | 20. | Number of lamps group replaced per year $(19 \cdot = \text{lamps/unit} \div 3)$ | | |
| | 21. | Number of interim spot replacements $(20 \cdot = \text{burn outs in GR interval})$ | | |
| | 21A. | Number of lamps spot replaced per year — No group relamping $(19 \cdot \text{lamps/unit} \div 2)$ | | |
| | 22. | Replacement lamp cost per year $(20 \text{ or } 21A \cdot \text{net lamp cost})$ | | |
| | 23. | Labor cost for group replacements $(20 \cdot \text{group labor rate/lamp})$ at $ _____ / lamp | | |
| | 24. | Labor cost for spot replacements $(21 \cdot \text{spot labor rate/lamp}$ at $ _____ / lamp | | |
| | 25. | Cost of cleaning per luminaire per year | | |
| | 26 | Annual energy cost per year $(5 \cdot 19 \cdot \text{¢/kWH} \div 100\,000$ at _____ ¢ kWH | | |
| | 27. | Total annual operating cost per luminaire $(22 + 23 + 24 + 25 + 26)$ | | |
| | 28. | Relative annual operating cost for equal maintained footcandles $(27 \cdot 11$ of system compared $\div$ 27 of base system) | | |
| **Total** | 29. | Total annual cost — owning and operating — per luminaire $(17 + 27)$ | | |
| | 30. | Relative total annual cost for equal maintained footcandles $(29 \cdot 11$ of system compared $\div$ 29 of base system) | | |

Figure **6-2**  A typical lighting cost analysis form

(2) **Life-Cycle Costing Analysis** – Life-Cycle Costing (LCC) analysis is one technique which allows consideration of all the relevant economic consequences of design decisions in terms of money spent today (present cost) or money required during the life of the structure (annual cost). So LCC is the evaluation of a proposal over a reasonable time period considering all pertinent costs and the time value of the money. The evaluation can take the form of a present value analysis or uniform annual cost analysis. More sophisticated analysis would include sinking fund and rate of return on extra investment. This method of analysis is not intended to be a detailed study of various systems. It is to be used by practicing engineers as a guide for comparing the advantages of alternative design cases.

A more detailed analysis taking into account other items such as future costs could be performed. Figure 6-3 shows an outline of one method of determining costs, called "Life-Cycle Cost Analysis".

## Examples of Cost Analysis

A comprehensive cost analysis comparing an existing mercury lighting system with four alternative systems was made for an industrial plant. It served as a basis for selecting a best system which is not only more energy effective, but delivers better quality illumination at the same time. Table 6-1 exhibits the comparative cost analysis. Since the study was done several years ago, the material and labor cost figures contained therein may not be up-to-date, however, the relative cost rankings should still hold true for evaluation purposes. Figure 6-4 shows an economic analysis for two different lighting systems in a 30 ft by 30 ft room, to illustrate the application of the life-cycle costing method; it is by no means to show the benefits of one system over another.

Life cycle cost analysis for _____ ft$^2$ _____.

Luminaire _____ Luminaire _____<br>
Layout _____ Layout _____

**A. Lighting and air conditioning installed costs (initial)**

1. Luminaire installed costs: luminaire, lamps,
   material, labor ................................................ $_____ $_____
2. Total kW lighting: ......................................... _____kW _____kW
3. Tons of air conditioning required for lighting:
   (3.41 × kW/12) ............................................. _____tons _____tons
4. First cost of air-conditioning machinery: @
   $_____/ton ................................................... $_____ $_____
5. Reduction of first cost of heating equipment: ... $_____ $_____
6. Other differential costs: .............................. $_____ $_____
   .................................................................. $_____ $_____
   .................................................................. $_____ $_____
   .................................................................. $_____ $_____
   .................................................................. $_____ $_____
7. Subtotal mechanical and electrical installed
   cost: ........................................................... $_____ $_____
8. Initial taxes: ............................................... $_____ $_____
9. Total costs: ................................................. $_____ (A1) $_____ (B1)
10. Installed cost per square foot: .................... $_____ $_____
11. Watts per square foot of lighting: ............... _____watts _____watts
12. Salvage (at $y$ years): ................................. $_____ (As) $_____ (Bs)

**B. Annual power and maintenance costs**

1. Lamps: burning hours × kW × $/kWh ........... $_____ $_____
2. Air conditioning: operation-hours × tons
   × kW/ton × $/kWh ......................................... $_____ $_____
3. Air conditioning maintenance: tons × $/ton ... $_____ $_____
4. Reduction in heating cost fuel used: _____ ... $_____ $_____
5. Reduced heating maintenance: MBtu × $/MBtu $_____ $_____
6. Other differential costs: .............................. $_____ $_____
   .................................................................. $_____ $_____
   .................................................................. $_____ $_____
   .................................................................. $_____ $_____
7. Cost of lamps: (No. of lamps
   _____ @ $_____/lamp per $N$) (Group relamp-
   ing every $N$ years, typically every one, two or
   three years, depending on burning schedule.) ... $_____ $_____
8. Cost of ballast replacement:
   (No. of ballasts _____ @ $_____/ballast per n)
   (n = number of years of ballast life.) ............. $_____ $_____
9. Luminaire washing cost: No. of
   luminaires _____ @ $_____ each. (Cost to
   wash one luminaire includes cost to replace or
   wash lamps.) ............................................... $_____ $_____
10. Annual insurance cost: ............................... $_____ $_____
11. Annual property tax cost: ............................ $_____ $_____
12. Total annual power and maintenance cost: ... $_____ (Ap) $_____ (Bp)
13. Cost per square foot: .................................. $_____ $_____

## Notes on analysis

**A.** 1. An estimate is prepared for material
and labor of the installation.

2. In the example that follows a 40-W
rapid-start lamp with ballasts loss is considered one 48-W load, and the 150-W HPS with
ballasts is considered 175-W.

4. First cost of machinery will vary from
$1000 to $2000/ton. Use the same value for
both systems.

Figure **6-3** Life cycle cost analysis form

Table **6-1** Comparative cost analysis for five different lighting systems

| | Existing 400 W Mercury System | 250 W Metal-Halide System | 400 W Metal-Halide System | 250 W HPS System | 400 W HPS System |
|---|---|---|---|---|---|
| Number of luminaires required | 108 | 97 | 68 | 70 | 42 |
| Luminaire spacing (square grid), ft | 9.62 | 10.15 | 12.13 | 11.95 | 15.43 |
| Initial lamp lumens per lamp | 22,500 | 20,500 | 34,000 | 30,000 | 50,000 |
| Lamp lumen depreciation factor | 0.78 | 0.83 | 0.75 | 0.90 | 0.90 |
| Estimated lamp life, hr | 24,000 | 10,000 | 15,000 | 24,000 | 24,000 |
| Average lamp replacements per year | 18 | 38.8 | 18.13 | 11.67 | 7 |
| Lamp net cost, dollars per lamp | 10.23 | 23.55 | 22.35 | 38.40 | 36.00 |
| Luminaire input watts | 450 | 285 | 460 | 300 | 475 |
| Average watts per sq. ft | 4.9 | 2.8 | 3.1 | 2.1 | 2.0 |
| Total connected load, kw | 48.6 | 27.65 | 31.28 | 21 | 19.95 |
| Luminaire per unit cost, dollars | -0- | 68 | 85 | 145 | 150 |
| Installation labor per unit, dollars | -0- | 36 | 36 | 36 | 36 |
| Installation cost summary | | | | | |
| Luminaire cost, dollars | -0- | 6,596.00 | 5,780.00 | 10,150.00 | 6,300.00 |
| Initial lamp cost, dollars | -0- | 2,284.35 | 1,519.80 | 2,688.00 | 1,512.00 |
| Installation labor cost, dollars | -0- | 3,492.00 | 2,448.00 | 2,520.00 | 1,512.00 |
| Total installation costs, dollars | -0- | 12,372.35 | 9,747.80 | 15,358.00 | 9,324.00 |
| Annual operating cost summary | | | | | |
| Lamp cost, dollars | 184.18 | 913.74 | 405.28 | 448.00 | 252.00 |
| Maintenance labor, dollars | 180.00 | 288.00 | 181.33 | 116.67 | 70.00 |
| Energy cost, dollars | 7,776.00 | 4,423.20 | 5,004.80 | 3,360.00 | 3,192.00 |
| Total annual operating cost, dollars | 8,140.14 | 5,724.94 | 5,591.41 | 3,924.67 | 3,514.00 |
| Relative operating cost, percent | 100.00 | 70.33 | 68.69 | 48.21 | 43.17 |
| Total annual cost summary | | | | | |
| Annual owning cost, dollars | -0- | 1,513.20 | 1,234.20 | 1,900.50 | 1,171.80 |
| Owning and operating cost, dollars | 8,140.14 | 7,238.14 | 6,825.61 | 5,825.17 | 4,685.80 |
| Relative owning and operating cost percent | 100.00 | 88.92 | 83.85 | 71.56 | 57.56 |
| Annual cost per fc per sq ft, dollars | 0.8107 | 0.7216 | 0.6832 | 0.5819 | 0.4681 |
| Lighting investment payback summary | | | | | |
| Annual operating cost, dollars | 8,140.14 | 5,724.94 | 5,591.41 | 3,924.67 | 3,514.00 |
| Operating cost savings, dollars | -0- | 2,415.20 | 2,548.73 | 4,215.47 | 4,626.14 |
| Total new investment, dollars | -0- | 12,372.35 | 9,747.80 | 15,358.00 | 9,324.00 |
| Simple investment payback interval, years | -0- | 5.12 | 3.82 | 3.64 | 2.02 |
| Simple return on investment, percent | -0- | 19.52 | 26.15 | 27.45 | 49.62 |
| Adjusted discounted investment payback interval, months | -0- | 90.4 | 60.7 | 57.0 | 28.3 |
| Summary of costs over next 20 years | | | | | |
| Net lamp costs, dollars | 3,682.80 | 17,132.62 | 7,345.70 | 7,616.00 | 4,284.00 |
| Lamp replacement labor costs (at $10 per lamp), dollars | 3,600.00 | 7,760.00 | 3,626.67 | 2,333.33 | 1,400.00 |
| Energy consumption, kwh | 3,888,000 | 2,211,600 | 2,502,400 | 1,680.000 | 1,596,000 |
| Total energy costs, dollars | 155,520.00 | 88,464.00 | 100,096.00 | 67,200.00 | 63,840.00 |
| Total initial costs, dollars | -0- | 12,372.35 | 9,747.80 | 15,358.00 | 9,324.00 |
| Total 20 year life-cycle costs, dollars | 162,802.80 | 125,728.97 | 120,816.17 | 92,507.33 | 78,848.00 |

• Basis: 10,000 sq ft manufacturing area illuminated to 100 fc, 4000 burning hours per year, effective electrical energy rate (including demand and other charges) 49 kwh, average dirt conditions, 20 year amortization at interest rate of 10 percent.

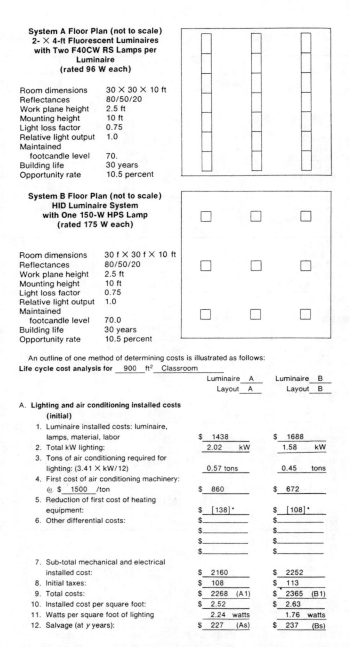

System A Floor Plan (not to scale)
2- × 4-ft Fluorescent Luminaires
with Two F40CW RS Lamps per
Luminaire
(rated 96 W each)

| | |
|---|---|
| Room dimensions | 30 × 30 × 10 ft |
| Reflectances | 80/50/20 |
| Work plane height | 2.5 ft |
| Mounting height | 10 ft |
| Light loss factor | 0.75 |
| Relative light output | 1.0 |
| Maintained footcandle level | 70. |
| Building life | 30 years |
| Opportunity rate | 10.5 percent |

System B Floor Plan (not to scale)
HID Luminaire System
with One 150-W HPS Lamp
(rated 175 W each)

| | |
|---|---|
| Room dimensions | 30 f × 30 f × 10 ft |
| Reflectances | 80/50/20 |
| Work plane height | 2.5 ft |
| Mounting height | 10 ft |
| Light loss factor | 0.75 |
| Relative light output | 1.0 |
| Maintained footcandle level | 70.0 |
| Building life | 30 years |
| Opportunity rate | 10.5 percent |

An outline of one method of determining costs is illustrated as follows:

**Life cycle cost analysis for** ___900___ ft² ___Classroom___

| | Luminaire A / Layout A | Luminaire B / Layout B |
|---|---|---|
| A. Lighting and air conditioning installed costs (initial) | | |
| 1. Luminaire installed costs: luminaire, lamps, material, labor | $ 1438 | $ 1688 |
| 2. Total kW lighting: | 2.02 kW | 1.58 kW |
| 3. Tons of air conditioning required for lighting: (3.41 × kW/12) | 0.57 tons | 0.45 tons |
| 4. First cost of air conditioning machinery: @ $ 1500 /ton | $ 860 | $ 672 |
| 5. Reduction of first cost of heating equipment: | $ [138]* | $ [108]* |
| 6. Other differential costs: | $_____ | $_____ |
| | $_____ | $_____ |
| | $_____ | $_____ |
| | $_____ | $_____ |
| | $_____ | $_____ |
| 7. Sub-total mechanical and electrical installed cost: | $ 2160 | $ 2252 |
| 8. Initial taxes: | $ 108 | $ 113 |
| 9. Total costs: | $ 2268 (A1) | $ 2365 (B1) |
| 10. Installed cost per square foot: | $ 2.52 | $ 2.63 |
| 11. Watts per square foot of lighting | 2.24 watts | 1.76 watts |
| 12. Salvage (at y years): | $ 227 (As) | $ 237 (Bs) |

Figure **6-4** Economic comparison analysis for two lighting systems for a classroom

**B. Annual power and maintenance costs**

1. Lamps: burning hours × kW × $/kWh     $ __314__       $ __246__
2. Air conditioning operation hours × tons
   × kW/Ton × $/kWh                     $ __72__       $ __56__
3. Air conditioning maintenance: tons × $/ton   $ __86__       $ __67__
4. Reduction in heating cost fuel used: coal   $ __[23]__*      $ __[18]__*
5. Reduced heating maintenance: MBtu
   × $/MBtu                                 $ __[14]__*      $ __[11]__*
6. Other differential costs:                 $_____      $_____
                                       $_____      $_____
                                       $_____      $_____
                                       $_____      $_____

7. Cost of lamps: (No. of lamps
   $\frac{42}{9}$ @ $ $\frac{1.08}{34.00}$ /lamp per N) $\frac{3 \text{ yrs.}}{6 \text{ yrs}}$
   (Group relamping every N years, typically
   every one, two or three years, depending on
   burning schedule.)                   $ __15__       $ __51__
8. Cost of ballast replacement: (No. of
   ballasts $\frac{21}{9}$ @ $\frac{12 \text{ yrs}}{12 \text{ yrs}}$ $ $\frac{12}{63}$ /ballast
   per n) (n = number of years of ballast life.)   $ __41__       $ __58__
   *Labor A = 0.8 hrs, B = 1.0 hrs;*
   *Rate = 14.50/hr*
9. Luminaire washing cost: No. of
   luminaires $\frac{21}{9}$ @ $ $\frac{7.25}{7.25}$ each. $\frac{1 \text{yr}}{1 \text{yr}}$
   (Cost to wash one luminaire includes cost
   to replace *or* wash lamps.)         $ __152__       $ __65__
   *Labor A & B = 0.5 hr; rate = 14.50/hr*
10. Annual insurance cost:                  $ __27__       $ __28__
11. Annual property tax cost:               $ __136__      $ __142__
12. Total annual power and maintenance cost: $ __806__ (Ap)   $ __684__ (Bp)
13. Cost per square foot:                   $ __0.90__      $ __0.76__

\* Bracket indicates negative values.

**PAYOUT (Example)**

(Sys. A) = ($2268 − $227) = $2041

(Sys. B) = ($2365 − $237) = $2128

$X_1$ = ($2128 − $2041) = $87

$X_2$ = ($806 − $684) = $122

$X_3 = \dfrac{122}{(.105 \times 87)} = 13.36$

$a = 1.105$

$b = \dfrac{13.36}{12.36} = 1.081$

$y = \dfrac{\ln 1.081}{\ln 1.105} = 0.78$ years before Sys. B begins to be the more economical system. □

Figure **6-4** Economic comparison analysis for two lighting systems for
a classroom (cont.)

## Bibliography

Chen, Kao, The Energy Oriented Economics of Lighting Systems, IEE Transactions on Industry Applications, Jan./Feb. 1976, pp. 62-68.

Chen, Kao, and Guerdan, E.R., Resource Benefits of Industrial Relighting Program, IEEE Transactions on Industry Applications, May/June 1979.

Chen, Kao, and Lally, William, Update: Fluorescent Lighting Economics, IEEE Transactions on Industry Applications, May/June 1983, pp. 328-333.

Chen, Kao, Design Applications in Lighting Retrofits, IEEE-IAS Annual Conference Proceedings, Oct. 1992.

Hart, A.L., Cutting Lighting Costs by Applying Energy Efficient Lighting Sources, Plant Engineering, Mar. 8, 1979.

Wellinghoff, Jon, Winning the Lighting Retrofit Game, EC&M, May 1993, pp. 65-76.

ANSI/IEEE 739-1984, Energy Conservation and Cost-Effective Planning in Industrial Facilities.

# Chapter 7

# Examples of New and Retrofitting Installations

## 7.1 Introduction

Chapter 6 has outlined various steps and methods to achieve energy effectiveness in the new and retrofit lighting systems. This chapter will cover examples of typical lighting installations which have demonstrated achieving the objectives of seeing tasks and energy effectiveness in those systems.

## 7.2 New Installations

(1) **Machine Shops**

Machining of metal parts consists of setting up and operating machines such as lathes, grinders, millers, shapers, and drill presses, bench work, and inspection of metal surfaces. The precision of such machine operation usually depends on the accuracy of the setup and careful use of the graduated feed-indicating dials rather than observation of the cutting tool or its path. The fundamental seeing problem is the discrimination of detail on plane or curved metallic surfaces.

The visibility of scribed marks depends on the characteristics of the surface, the orientation of the scribed mark, and the nature of the light source. Directional light produces good visibility of scribed marks on untreated cold-rolled steel if the marks are oriented for maximum visibility, such that the brightness of the source is reflected from the side of the scribed mark to the observer's eye. Unfortunately, this technique reduces the visibility of other scribed marks. Better average results are obtained with a large-area low-luminance source.

There is an obvious advantage in the use of large-area low-luminance source for most visual tasks in the machining of metal parts. The ideal general lighting system is one having a large indirect component. Both fluorescent and HID sources can be used for general lighting; fluorescent luminaires in a grid pattern are usually preferred. High-reflectance room surfaces improve visual performance. Figure 7-1 shows the result of such a system, which provides a pleasant environment for die-making in a machine shop. Supplementary lighting is used to maintain close tolerances of die production. Green plants and modern wall treatment contribute to making a stimulating space.

Special types of luminaires are designed to illuminate three dimensional objects such as machinery, dials, spindles, presses, panels, and stacked materials. Light is contributed from many locations and distances to minimize shadow, reduce glare, and optimize visual comfort. This type of luminaire has proven to be satisfactory for lighting of machine shops. Figure 7-2 shows a different type of lighting

Figure **7-1** Fluorescent lighting for a machine shop with
supplementary lighting for die production

for another industrial plant machine shop, which is
lighted with 250 W metal halide luminaires with
emphasis on vertical surface illumination.

(2) **Control Rooms**

The control room is the nerve center of a power plant
or process plant and must be monitored continuously.
Lighting must be designed with special attention to
the comfort of the operator; direct or reflected glare
and veiling reflections must be minimized, and luminance ratio must be low. Along with ordinary office-
type seeing tasks, it is often necessary to read meters
10 to 15 ft away.

Figure **7-2** Metal halide lighting for an industrial plant machine shop

Although the practice is not standardized, most control room lighting involves one of two general categories: diffuse lighting or directional lighting. Diffuse lighting may be from low-luminance, luminous indirect lighting equipment, solid luminous plastic ceilings, or louvered ceilings. Directional lighting may be from recessed troffers that follow the general contour of the control board.

A basic conflict exists in trying to light a control room, since some of the tasks are made more visible under reduced illumination, while other tasks require significantly higher levels of illumination. Fortunately, most of the tasks enhanced by low light levels are not located in the same area where tasks requiring high illumination are to be found. Therefore, the lighting system can be modified to give nonuniform distribution of light within the room.

The optimum distribution of illumination for each control room must be determined on a case-by-case basis since the equipment arrangement will vary in each installation. It may also be found necessary to use several different types of louvers, lenses, and diffusers, sometimes in combination with one another, to achieve the optimum distribution for each particular room.

(3) **VDT Rooms**

Currently, there is a proliferation of VDT (visual display terminals) in all areas of industry. Lighting for these areas needs special attention. The lighting design should provide for a good visual performance, comfort, and creation of a psychologically and aesthetically pleasing environment. In specific terms, the lighting design should still limit both direct and indirect glare, and control luminance in both the immediate task area, and within the dynamic field of view.

a) *Illuminance* – Illuminance values can be determined from the IES Handbook. Most VDTs are self-luminous, and require virtually no illuminance for task contrast. However, one must consider adjacent tasks, comfort, and psychological well being. As many adjacent tasks will fall into category "D", illuminance values of 20-30-50 fc are most likely to be appropriate. Typically, 20-40 fc of general illumination should be the maximum, with task lighting supplementing where necessary, with quality (no veiling reflection), controlled brightness luminaires.

b) *Luminance* – Luminance is illuminance times reflectivity. Current practice suggests that lumi-

nance ratios are important to good visual performance and comfort. Within the "Task Surround",
luminance values should not be greater than 3
times that of the task, or less than 1/3 the level
of the task. These standards may be conservative
but provide a useful guideline. In measuring a
variety of VDTs, screen luminance typically
ranges from 5-25 fL, with 15 fL a reasonable
average.

c) *Luminaires and Indirect Glare* – The most common complaint of office lighting today is indirect
glare on VDT screens. All types of indirect glare
are a result of brightness in the "Offending Zone"
(see Figure 3-1).

d) *Light Fixture Location* – The best place to locate
fixtures is to the side of the VDT; such positioning will minimize reflection of the fixture on the
screen. Figure 7-3 shows such arrangement.
Optimum placement of fixture for a particular
VDT might cause problems for other VDT operators. So conscientious planning is required for
both the positioning of light fixtures and the
placement of work stations. The use of three-
lamp fixture should be considered because such
luminaires are more energy efficiency, and they
permit three levels of switching. Another basic
method to eliminate fixture reflections in the face
of the VDTs is to equip the fixture with specular
parabolic wedge louver, with an absolute cutoff of
45 degrees; so lamp images will be reflected
below the viewing angle.

## (4) **Manufacturing Areas**

Fluorescent lighting was widely used for industrial
operations prior to the energy crunch in the mid-70s.
During the last decade, HID (particularly the metal

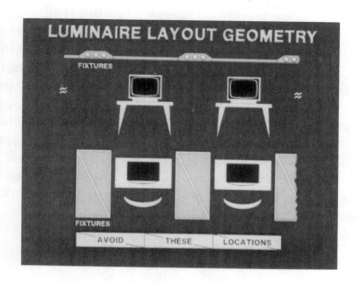

Figure **7-3** Well-shielded luminaires placed to the sides of the work station

halide and high-pressure sodium) lighting has been becoming more applicable in industrial plants.

In high-bay areas, 1000 W HPS luminaires, which provide some uplight component through open and ventilated reflectors are used in an approximate spacing-to-mounting height of 1:1. For low-bay areas, 400 W luminaires with about a 16 ft mounting height provide even candlepower distribution of light without glare using a faceted aluminum reflector and a polycarbonate lens having refractor prism elements. The optical assembly is totally enclosed, gasketed, and filtered to keep contaminants from infiltrating to the lamp and reflector or inside the lens. Since the lens prisms redirect some of the high-angle lumen output toward lower angles, these industrial units provide adequate horizontal beam spread and vertical surface illumination.

In general, the spacing and mounting-height arrangements of both low- and high-bay units provide approximately 60 fc horizontal maintained on a typical work surface. All the fixtures are wired in a checkerboard pattern on 480 V three-phase circuits. The overall light level can be reduced 50% in a given zone by switching off half the circuits in that zone from a convenient central point. Thus energy savings can be achieved during reduced production activity. Each fixture is powered through a fused plug and cord assembly which permits easy replacement of the unit for repair.

(5) **Warehouses**

Designing a lighting system for a warehouse might appear to be relatively simple. Install enough luminaires to deliver enough light onto the stored material to permit accurate selection, and in the area for operators to see safely where they are going. Over the years, a few rules of thumb have evolved as offshoots of manufacturing area lighting practice. However, in recent years the concept of storage has changed. Warehouses are larger, stacks are higher and deeper, and operations have become more automatic and computer controlled. In designing lighting for most new warehouses, illuminating engineers find their job comparable to designing a system for a long, tall, narrow room for which reflectances are unknown and in which luminaires can be placed only in high, inaccessible places.

In the past (before energy crunch years), fluorescent lighting was commonly used to illuminate vertical surfaces. Figure 7-4 shows high-output fluorescent luminaires operating on 277 V circuits installed at a mounting height of 40 ft in a modern warehouse (built in 1966). It delivered satisfactory vertical illumination, especially near the top of the stacks.

Figure **7-4** High output fluorescent lighting for warehouse aisles

Today, higher mounting heights have prompted the use of HID lighting equipment. Because the light is emitted from an optically small area, a HID lamp permits finer optical control than do line source fluorescent lamps; luminaires can be designed with better directional characteristics than those for line source.

In general, several factors are unique in warehouse lighting. First, there is the necessity of seeing on a vertical surface rather than on a horizontal plane. Second, the warehouse aisle is similar to a narrow, long room with high walls, where inter-reflectances affect the result to a significant degree. Third, the type and amount of material in the warehouse aisles are subject to unpredictable fluctuation.

Recent studies on warehouse aisle lighting have pointed out that luminaires with maximum uplight provide superior vertical as well as horizontal illumination at all locations for several aisle widths. The uplight would be reflected from the ceiling, which would be advantageous in a warehouse aisle setting with narrow aisles and high mounting heights. The study concluded that the amount of uplight appears to be an important factor and that luminaire design appears to be one of the most critical elements in the production of vertical illumination. Figure 7-5 shows an empty warehouse lighted with 400 W HPS luminaires, resulting in good vertical illumination. The HPS lamps were also chosen because of the cost factor.

Illuminating Engineering Society (IES) has recently published a "Design Guide for Warehouse Lighting". Table 7-1 lists recommended maintained illuminance values for the various storage, shipping, and receiving spaces in warehouses. Illuminance ranges in fc or lux are to be used for design purposes as targeted maintained values. Proper use of this table in conjunction with Tables 2-3 and 2-4 can assist illuminating engineers to determine a most appropriate illuminance value required for the design.

Figure **7-5** 400 W HPS lighting for an unoccupied warehouse

Table **7-1** Recommended Illuminance Categories for Warehouse Lighting

| Type of Activity | Illuminance Category | Lux | FC |
|---|---|---|---|
| Inactive or rough and bulky storage | B | 50-75-100 | 5-7.5-10 |
| Active or medium storage | C | 100-150-200 | 10-15-20 |
| Small Item Storage | D | 200-300-500 | 20-30-50 |
| Shipping & Receiving | D | 200-300-500 | 20-30-50 |
| TV Surveillance | Contact Manufacturer of TV Camera | | |

Notes: Recommended footcandles are based on maintained average illuminance levels and should include the proper total light loss factor.

Recommended footcandles are for both vertical and horizontal surface.

Uniformity of illumination measured in terms of average/minimum should not exceed 10:1 on vertical surfaces and 3:1 on horizontal surfaces.

(6) **Engineering Offices, Conference Rooms, and Plant Hospital Rooms**

   a) *Engineering Offices* – Visual requirements for engineering and drafting offices demand high quality illumination since discrimination of fine detail is frequently needed for extended periods. Harsh directional shadows from drawing instruments may reduce efficiency. Illumination systems that avoid reflections are most important in providing maximum contrast. Studies indicate that ideal locations for lighting fixtures were at least 24 in. from either side of the drafting table. In general, the recessed 2 x 4 ft parabolic fluorescent troffers with 4 in. deep cells and four lamps are suited for such applications. This type of lighting system can produce virtually glare-free and adequate illumination at the center of the drafting tables.

   b) *Conference Rooms* – The diversity of work to be performed in the conference room requires that the illumination should be flexible and the entire room should be comfortable and pleasant. The general lighting is provided by the recessed fluorescent in the 2 x 4 grid-type ceiling. Specially chosen reflectors provide high-level illumination without glare. Incandescent (250 W PAR 38) downlights with concentric louvers are provided in the front stage and over the conference table. This, in combination with dimmer-controlled general lighting, provides flexibility to create a changing environment, as well as eye comfort for all seeing tasks. Figure 7-6 shows a handsomely decorated functional conference room.

Figure **7-6** Combination of fluorescent and incandescent lighting for a functional engineering conference room

c) *Plant Hospital Rooms* – A plant hospital room is used for emergency and/or simple treatments only. It is staffed with a nurse and a part-time doctor. In the hospital room a moderately high illuminance is required for examination.

Ideally, a 50 fc level for local examination is recommended. It is advantageous to use three-lamp or four-lamp fluorescent luminaires for the general illumination, switched or dimmed to allow reduction of the light level during certain procedures. This type of room can be better served with general illumination supplemented by a portable or fixed examination light. The general light source should be of the color-improved type. Figure 7-7 illustrates such a room.

Figure **7-7** Level controlled fluorescent lighting for a plant hospital room

## 7.3  Retrofitting Installations

### 7.3.1  The Importance of Energy Effectiveness in Retrofitting

It has long been recognized that energy effective illumination design has a tremendous impact on the costs.
Here the costs include not only the initial cost of the
installation, but also the operating and maintenance
costs. Often energy effective design will result in an
increase of the initial cost over the traditional approach.
However, often when lighting quality is improved as a
result of successful retrofitting, higher worker productivity as well as energy/cost savings can be achieved.
Therefore, the engineer must not fail to recognize that
savings can result from improved productivity, few errors

and rejects, increased safety and security, lower liability insurance premiums, and increased retail sales, etc. In many instances, the value to be derived from better design and hence the better performance of the retro-fitting system is worth ten or more times what would be realized from even the most significant reduction in energy consumption. So the engineer must not lose sight of the importance of the system performance to the energy savings in the initial stage of the project.

## 7.3.2 Examples of Retrofitting Installations

The following examples are intended to illustrate several excellent projects which have achieved not only energy/cost savings, but improved employee's visual performance as well. One example is purposely included to illustrate a total failure due to a non-professional's recommendation, which treated the retrofit project as a replacing game focusing on the energy savings alone without studying the workers' visual requirements, and resulted in an unsalvageable installation.

(1) **Retrofit Options for a Mercury Lighting System**
The high operating cost of a mercury lighting system is second only to that of an incandescent system. For maximum savings, a mercury system can be replaced with a high pressure sodium system, or if color discrimination is important, a metal halide system. Two basic retrofit options can be applied to a mercury system. The first option is to replace the luminaire ballasts with either metal halide (MH) or high pressure sodium (HPS) units. The second option is to replace with special MH or HPS lamps designed to operate on mercury ballasts. Replacing ballasts require a larger cash expenditure, but the long-term benefits are greater. Although retrofit lamps produce

substantial savings, they generally cost more and have a shorter life, or lower efficacy or lumen maintenance than those of standard MH or HPS lamps operated on companion ballasts.

At the present time, retrofit HPS lamps in wattages of 150, 215, and 360 W can be used on mercury systems using 175, 250, and 400 W lamps respectively. One manufacturer also offers 880 W HPS lamps for use in existing 1000 W mercury luminaire. Retrofit MH lamps are available in 400- and 1000 W versions, which are intended for use with CW/CWA type mercury ballasts. The 1000 W MH can reduce wattage per luminaire by either 35 W or 85 W, depending on the type of ballast it is used with. A new addition is the 325 W MH lamp which saves about 70 W per luminaire while delivering 40% more light than 400 W mercury lamp it replaces.

**Fixture for Fixture Replacement** – This pattern of relighting represents one of the simplest conditions. A typical case for such a system is to replace 1000 W mercury lights with 400 W HPS on a one-for-one basis. Usually, the spacing-to-mounting height ratio is suitable for the new light source. The existing wire sizes are usually adequate and connections can easily be made. The overall installation costs of the retrofit program will be low, and the ROI will be high. The lighting level will stay practically the same as before. However, the energy savings realized in this retrofit program amount to 60% of the original consumption. Figure 7-8 shows a typical industrial plant relighted in this nature. On the left, one hundred seventy eight 1000 W mercury luminaires were used to illuminate the plant; on the right, the same number of 400 W HPS new luminaires are in place. The ROI in this case is 58%.

Figure **7-8** A typical industrial high-bay retrofit project – 400 W HPS replacing 1000 W mercury

## (2) New California Post Office Building

The building was not designed to comply with Title 24 because it was a federal building. Negotiating with the utility the building was able to secure a rebate for daylighting under the utility's Title 24 program, which pays for energy efficiency measures that beat the state building code by at least 1%.

The daylighting system's skylights, roof closures, and controls cost $540,000 and are expected to cut 3,334,000 kWh in energy consumption annually. This is expected to save $147,000 from the facility's utility bills. With a rebate, the payback for daylighting will be about 2.7 years.

The daylighting system is used to supply lighting for a single-story workroom. The building remaining area over two floors does not include daylighting controls for its electronically ballasted T8 lighting.

The workroom daylighting system includes 280 skylights covering 3.4% of the workroom roof. Each skylight measures about 4 x 8 ft and includes an acrylic lens with ultraviolet light stabilizers. They have a viable light transmittance of 72% and a shading coefficient of 66%. The system also includes a master time control to control 1650 250-W HPS lamps. The luminaires are configured in a diamond pattern in each of seven zones.

Based on photocell readings, the controller can turn off one to three lamps in a diamond when ambient light is sufficient to meet set points. The rest of the building lighting consists of 32 W T8 fluorescent lamps with electronic ballasts. The installation of 9782 T8 lamps and 3820 two-, thee-, and four lamp ballasts is expected to cut additional 515,000 kWh from the building's annual electric energy consumption.

## (3) A Telephone Company's Project

The project involved replacing four 40WT12 lamps and two magnetic ballasts in each of 1600 fixtures with two 32WT8 lamps, a single electronic ballast, and a reflector. This retrofit reduced demand from 192 to 62 W per fixture to a total demand reduction of approximately 208 kW. Annual energy consumption consequently fell by 865,000 kWh, based on 4160 hours of operation per year. The installation will cut operating costs at the 163,000 sq. ft. buildings by an estimated $63,000 a year. A projected $52,000 in annual utility bill reduction based on average savings of $4,300 realized between February and June. An additional $5,600 in maintenance and equipment cost savings comes from reduced ballast failure and fewer lamp failures. Finally, $5,700 savings comes from the estimated 41 ton reduction in cooling demand stemming from the delamping.

The project's 1.7 year simple payback is further shortened by a $41,000 incentive paid by the utility. This incentive brought the project's payback to just under 13 months.

## (4) A Machine Shop

This example illustrates the consequence of playing a replacing game to save energy while ignoring the recommended practice in lighting design.

The existing machine shop has a 160 x 200 ft production area in which medium to fine machining work is performed. The existing lighting system consists of stem-mounted 8 ft high-output (HO) fluorescent fixtures in continuous rows on 12 ft centers as shown in Figure 7-9. These fixtures provided a uniform horizontal maintained illuminance of 70 fc. The plant managers were ill advised by a non-professional consultant that lighting energy costs could be reduced by 50% through the use of HPS lighting. Specifically, individual 400 W HPS fixtures, stem mounted at 25 ft above the floor were installed. Immediately after, employees began complaining about poor visibility at their work stations.

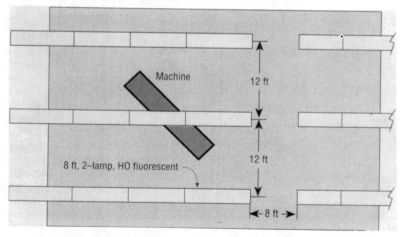

Figure **7-9** A machine shop layout with fluorescent lighting

After analyzing the data obtained from the site, two important design practices were found to be violated. First, the designed illuminance level was far below that recommended by IES for the type of work being performed. Second, the photometric distribution pattern of the luminaire was not compatible with the fixture layout within the area. The first error was supported by the light meter readings, which ranges from 10 to 40 fc on the horizontal work plane, far below the 70 fc level of the original fluorescent lighting and that called for by IES. The second error was established after studying the HPS fixture manufacturer's data sheet. In this case, the HPS luminaire has a hemispherical reflector and a prismatic refractor lens enclosing the lower part of the reflector's housing. The distribution of this fixture concentrates the greatest amount of light within a 25 degree angle on either side of the nadir. This fixture is termed a sharp-cutoff unit. Consequently, most of the lumen output falls only within a 20 ft diameter circular area of the floor for this mounting height, and only directly below the fixture. An ideal application for this type of fixture would require a mounting height of 40 ft or more.

An even more revealing error was uncovered by carrying out a point-by-point computation. Figure 7-10 shows a four-fixture work station layout, typical of many of the work stations. Table 7-2 shows the calculated light contributions of each of the four fixtures. Note that Fixture A contributes over half of the light. Problem areas surface in the form of shadows created by the operator when blocking light from fixtures A & D. Machine positioning also creates problems because fixtures B & C contribute relatively little light to the task area.

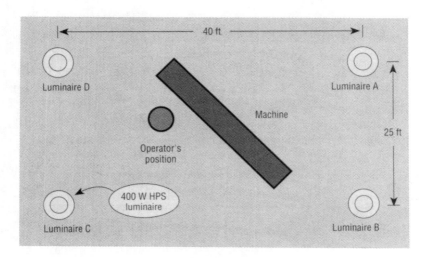

Figure **7-10** A typical four-fixture work station layout

Table **7-2** Light Contributions from HPS fixtures originally installed

| Luminaire | angle from nadir | Candle-power (cd) | Horizontal footcandles | % contribution |
|-----------|------------------|-------------------|------------------------|----------------|
| A | 17 | 16000 | 18.79 | 53 |
| B | 39 | 6800 | 4.29 | 12 |
| C | 40 | 6800 | 4.11 | 12 |
| D | 39 | 10500 | 8.32 | 23 |
| | | Totals | 35.51 | 100 |

The light distribution of a suggested replacement HPS luminaire indicates that it would contribute light at larger angles from the nadir and would be better suited to the 25 ft mounting height application. If this type of luminaire were specified, all four of the fixtures would have a more even share of light contribution as shown in Table 7-3. Also note that the illuminance on the task is increased by 20%. However, the average maintained illuminance level would still

be below the recommended 70 fc. So the final solution is to reinstall the original Fluorescent HO luminaires, but retrofit them with electronic ballasts to
obtain 25% reduction in energy costs. In addition,
this system will produce fewer flickers as reported in
its operation.

Table **7-3** Light Contributions from suggested HPS fixtures to be
mounted in same locations

| Luminaire | angle from nadir | Candle-power (cd) | Horizontal footcandles | % contribution |
|-----------|------------------|-------------------|------------------------|----------------|
| A | 17 | 10200 | 11.98 | 29 |
| B | 39 | 14000 | 8.83 | 21 |
| C | 40 | 14100 | 8.51 | 20 |
| D | 33 | 16200 | 12.83 | 30 |
|   |    | Totals | 42.15 | 100 |

# Bibliography

Chen, Kao, Design Applications in Lighting Retrofits, IEEE-IAS Annual Conference Proceedings, Oct. 1992.

Ferzacca, Nicola D., Avoid the Pitfalls of Lighting Retrofit, EC & M, April 1993, pp. 59-62.

Rowe, Gorden D., Solving Lighting Problem in VDT Areas, Plant Engineering, Dec. 27, 1984.

VDT Lighting, IES RP-24-1989, Illuminating Engineering Society of North America, New York, N.Y.

Design Guide for Warehouse Lighting, IES DG-2-1992, Illuminating Engineering Society of North America, New York, N.Y.

# Chapter 8

# Floodlighting Design Fundamentals

## 8.1 Function of Floodlighting

Floodlighting a building, a monument, or a structure is an effective way to identify the object at night and thereby calling attention to it and to its owner. Thanks to the recently developed light sources, luminaires, and techniques, lighting effects can be tailored to the type of building or structure and the significance the owner wants to give it. However, the equipment and techniques must be used intelligently and imaginatively, for it is essential that the building or structure's form, beauty, and architectural identity be neither disturbed nor obscured.

In general, floodlighting should achieve certain objectives:

(1) the structure surface should have such a brightness that it appears in perspective when viewed from a distance. Shadows cast should look like those cast by the sun; they should not destroy the basic form and depth of the structure's architecture. Walls and

other flat surfaces should be illuminated to a level that reveals their texture and the character of the architectural design.

(2) the structure should be identified with the area about it by illuminating sufficient surrounding area; that is, it should not appear suspended but rather oriented with adjacent grounds, slopes, and plazas.

## 8.2 Basic Floodlighting Effects

(1) **Flat Lighting** – this is uniform illumination of a structure. It creates few highlights and shadows and little modeling, but it can be the most economical kind because installation is usually simple and little of the light pattern misses the building. Luminaires can be mounted on the ground, on poles, or on the roofs of adjacent buildings or buildings across the street.

(2) **Grazing** – dramatically expresses the character of a building by producing strong highlights and shadows. It can be achieved by mounting floodlights close to the facade, so this method is often used when mounting space is limited. The best light source for tall buildings is a high intensity discharge (HID) lamp with its arc tube along the axis of a concentrating specular reflector.

(3) **Sparkle or Glitter** – this effect is achieved with exposed lamps. It can complement modern architecture with its emphasis on line and plane. The lamp size required for a sparkle pattern depends on the brightness of the area and the effect desired.

(4) **Patterns** – this is used to emphasize or subdue adjacent architectural elements, strengthen design concepts, or increase the attraction of an otherwise plain surface. The key to success in nonuniform lighting is to create the impression that the effect was planned.

(5) **Color** – it can supplement the increasing use of bold colors in modern construction, both in general flood-lighting and as a means of establishing highlights and focal points. This can be achieved either by use of color filters or by utilizing the inherent color differences among the light sources. Incandescent lighting produces a natural look; clear mercury lighting tends to cast a slight greenish color on natural colors; fluorescent lighting strengthens white or light blue colors; and sodium lighting is rich in amber color and very effective in adding warmth.

## 8.3. Selecting the Light Source

The basic categories of light sources are incandescent, fluorescent, and HID lamps. Table 8-1 compares the costs, life and characteristics, color, size, and aesthetic achievement of the various light sources.

Table **8-1** Comparison of light sources for floodlighting applications

| | Incandescent | | | Mercury | | |
|---|---|---|---|---|---|---|
| | Standard | Quartz-Iodine | Fluorescent | Standard | Metallic Additive | High-Pressure Sodium |
| Initial Cost | Low | Low | Higher | Higher | Higher | Higher |
| Annual Operating Cost | Medium | Medium | Low | Low | Low | Low |
| Service Life | Fair | Fair | Good | Very good | Good | Good |
| Color Definition | Good | Very good | Fair | Fair | Good | Good |
| Beam Control | Very good | Good | Poor | Fair | Good | Good |
| Cold Weather Operation | Very good | Very good | Fair | Good | Good | Good |
| Long Range Projection (narrow beam) | Very good | Fair | Poor | Fair | Fair | Fair |
| Medium Range Projection | Good | Good | Fair | Good | Good | Good |
| Lumen Output | Fair | Fair | Fair | Good | Very good | Best |

Incandescent lamps are perhaps the most useful and versatile floodlight sources. Their light can be directed easily by lenses and reflectors in beams of the desired shape, and the color of their light is accepted as "white". Efficacy usually is about 20 lm/W.

Tungsten halogen lamps, the new incandescent source, have efficacies of about 25 lm/W. They contain halogen that continually removes vaporized tungsten deposits from the quartz envelope and redeposits it on the filament; consequently their light output remains almost constant over time instead of diminishing as a result of tungsten depositing on the envelope. The lamp used for building floodlighting are usually about the size and shape of pencils. More floodlights developed for these linear light sources produce rectangular beam patterns, which are highly efficient for many building floodlighting applications.

Projector lamps are developed for particular needs. The 6-V 120 W PAR64 incandescent lamp produces a thin beam that is very effective for floodlighting tall buildings, columns, steeples, water towers, and the like. Its beam spread of 4-1/2 degree in one plane by 7 degree in the other is achieved by masking critical areas of the reflector to prevent refocusing of light.

Fluorescent lamps are lower in brightness than the other sources, but more efficient than most. Fluorescent lamps require a large specular reflector for precise control of light, but even with such a reflector, control is limited to the light perpendicular to the length of the lamp. Fluorescent lamps are sensitive to temperature both in starting and operating, although outdoor type ballasts insure reliable starting down to -20 degree. Regardless of ballast type, light output is reduced when the lamp is exposed to low temperature and moving air.

Mercury vapor lamps are almost as efficient as fluorescent lamps, and somewhat more compact. Color rendition in general is inferior to that of incandescent lighting, although some mercury lamps have relatively good color rendition.

Metal halide lamps furnish approximately 75 to 100 lm/W of white light. Advantages over mercury lamps are:

(1) good color without the use of phosphor.
(2) high initial light output.

A characteristic of metal halide lamp is some variation in color uniformity from lamp to lamp. This is influenced by fluctuations in the voltage, ballast output characteristics, luminaire design, and ambient temperature. However, a metal halide light source produces a wide range of less subtle colors including yellow, red, green, and blue.

High-Pressure Sodium lamps have high efficacy of 100-130 lm/W of white light with a yellow/orange tone, which provides a rich warm amber color that serves the building material well. The appearance of natural surfaces lighted with high-pressure sodium lamps is similar to their appearance under warm white fluorescent or low-wattage incandescent lamps, but colors at the "cool" end of the spectrum are substantially greyed down.

## 8.4 Choosing A Luminaire

The first step in determining the type, number, and size of floodlight luminaires required to light a building is to choose a tentative floodlight on the basis of type of light source, shape and size of beam (round or rectangular; wide, medium, or narrow), and wattage or light output (beam lumens) of the source. In general, if a single requirement must be met, the illuminating engineer simply selects the lamp and luminaire best suited for the job. Where there is no clear-cut requirement, he compares the various lamp and luminaire characteristics and weighs the importance of each.

Table **8-2** Floodlight luminaire types

| Beam Spread (degrees) | NEMA Type | Minimum Efficiency (percent) | | | | Fluorescent |
|---|---|---|---|---|---|---|
| | | Incandescent | | Mercury | | |
| | | Effective Reflector Area (sq. in.) | | | | |
| | | Under 227 | Over 227 | Under 227 | Over 227 | Any |
| 10 to 18 | 1 | 34 | 35 | — | — | 20 |
| 18 to 29 | 2 | 36 | 36 | 22 | 30 | 25 |
| 29 to 46 | 3 | 39 | 45 | 24 | 34 | 35 |
| 46 to 70 | 4 | 42 | 50 | 35 | 38 | 42 |
| 70 to 100 | 5 | 46 | 50 | 38 | 42 | 50 |
| 100 to 130 | 6 | — | — | 42 | 46 | 55 |
| 130 and up | 7 | — | — | 46 | 50 | 55 |

Source: National Electrical Manufacturers' Association. Asymmetrical-beam floodlights may be designated by a combination type designation which indicates horizontal and vertical beam spreads in that order; e.g., a floodlight with a horizontal beam spread of 75 degrees (Type 5) and vertical spread of 35 degrees (Type 3) would be designated as a Type 5X3 floodlight.

If there are more than one suitable light sources, an economic study should be made to determine which would be the best choice for a number of years of service. The comparison of light sources in Table 8-1 can be effectively used as a quick selector. With the light source chosen, a luminaire can also be selected. Floodlight luminaires are usually divided into seven types on the basis of beam spread. In Table 8-2 it is noted that beam efficiencies vary with the type of beam and lamp, as shown. Two popular types of floodlights are shown in Figures 8-1 and 8-2.

Figure **8-1**  A typical floodlight for 400 W HID lamps
(courtesy of Westinghouse Electric Corp.)

## 8.5  Design Procedures

### (1)  Determine the Decorative Effect Desired

The first step in floodlighting design is to establish
the effect desired, or more accurately, to investigate
the effects possible. The daylight appearance may be
helpful. Daylighting is usually a combination of strong-
ly directional sunlight and diffuse sky light. The color
of the former is warm, while that of the latter is cool.
Shadows are never "black", but simply less bright and
bluer. Daylighting varies continually with the time of

Figure **8-2**  A typical floodlight for 1000 W HID lamps.

day and year and with the weather. However, floodlighting is highly controllable and can therefore be
utilized to present the building in a continuously
favorable aspect.

## (2) **Determine the Level of Illumination**

In Table 8-3 are listed the recommended illumination levels for many floodlighting applications. The illuminance levels should not fall below these values at any time in the maintenance cycle; therefore, an allowance for reasonable depreciation must be made in the design. In floodlighting buildings, monuments, and the like, the reflectance of the surface and the brightness of the surroundings must be considered in determining the amount of light necessary. If a building is located in an area that is normally crowded, it would be advisable to reduce the brightness on the lower portion of the building to prevent annoyance to pedestrians and motorists.

Table **8-3** Recommended levels of illumination for floodlighting applications

| | Recommended Footcandles (Minimum At Any Time) | | Recommended Footcandles (Minimum At Any Time) |
|---|---|---|---|
| **Building—** | | Parking Lots | 5 |
| General Construction | 10 | Self-Parking | 1 |
| Excavation Work | 2 | Attendant Parking | 2 |
| | | | |
| **Building Exteriors and Monuments, Floodlighted—** | | Piers, Freight and Passenger | 20 |
| Bright Surroundings— | | Prison Yards | 5 |
| Light Surfaces | 15 | | |
| Dark Surfaces | 50 | Quarries | 5 |
| Dark Surroundings— | | | |
| Light Surfaces | 5 | **Railroad Yards—Classification** | |
| Dark Surfaces | 20 | Switch Points | 2 |
| | | Body of Yard | 1 |
| **Bulletins and Poster Boards—** (Water Tanks or Stacks With Advertising Messages, Flags) | | **Service Stations (At Grade)—** | |
| | | Light Surroundings— | |
| Bright Surroundings— | | Approach | 3 |
| Light Surfaces | 50 | Pump Island Area | 30 |
| Dark Surfaces | 100 | Service Areas | 7 |
| Dark Surroundings— | | Dark Surroundings— | |
| Light Surfaces | 20 | Approach | 1.5 |
| Dark Surfaces | 50 | Pump Island Area | 20 |
| | | Service Area | 3 |
| Coal Yards (Protective) | 0.2 | | |
| | | **Shipyards—** | |
| Dredging | 2 | General | 5 |
| | | Ways | 10 |
| Loading Platforms | 20 | Fabrication Area | 30 |
| | | | |
| Lumber Yards | 1 | Storage Yards, Active | 20 |

(3) **Choose Proper Spread**

Table 8-2 shows that the floodlight equipment is divided into seven types according to beam spread, which is defined as the angle between the two directions in which the candlepower is 10% of the maximum candlepower at or near the center of the beam.

Beam efficiency is defined as the percentage of the beam lumens bear to the lamp lumens, the beam lumens being the lumens contained within the beam spread.

Although the choice of beam spread for a particular application depends on individual circumstances, the following general principles apply:

a) The greater the distance from the floodlight to the area to be lighted, the narrower the beam spread desired.

b) Since by definition the candlepower at the edge of a floodlight beam is 10% of the candlepower near the center of the beam, the illuminance level at the edge of the beam is one-tenth or less of that at the center. To achieve reasonable uniformity of illumination, the beams of individual floodlight must overlap each other as well as the edge of the surface to be lighted.

c) The percentage of beam lumens falling outside the area to be lighted is usually lower with narrow-beam units than with wide-beam units. Thus narrow-beam floodlights are preferable where they will provide the necessary degree of uniformity of illumination and the proper footcandles level.

(4) **Determine the Coefficient of Beam Utilization**

To determine the number of floodlights that will be required to produce a specified level of illuminance in a given situation, it is necessary to know the number of lumens in the beam of the floodlight and the percentage of the beam lumens striking the area to be lighted. The beam lumens may be obtained from manufacturers' catalogs. The ratio of lumens striking the floodlighted surface to the beam lumens is called the coefficient of beam utilization (CBU). When an area is uniformly lighted, the average CBU of the floodlights in the installation is always less than 1.0.

The coefficient of beam utilization for any individual floodlight will depend on its location, the point at which it is aimed, and the distribution of light within its beam. In general, the average CBU of all the floodlights in an installation should fall within the range 0.60 to 0.90. If less than 60% of the beam lumens are utilized, it is an indication that a more economical lighting plan should be possible by using different locations or narrower-beam floodlights. On the other hand, if the CBU is over 0.90, it is possible that the beam spread selected is too narrow and the resultant illumination will be spotty. An estimated CBU can be determined by experience or by making calculations for several potential aiming points and using the average figure thus obtained.

To make such calculations, the floodlighted area is superimposed on the photometric grid, and the ratio of the lumens inside this area to the total beam lumens is determined. All horizontal lines on a building appear as straight horizontal lines on the grid if the floodlight is so aimed that its beam axis is perpendicular to a horizontal line on the face of the building. All vertical lines except the one through the beam axis appear slightly curved. Figure 8-3 illustrates the superimposed method to determine the coefficient of beam utilization.

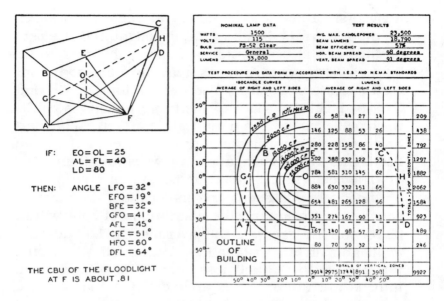

Figure **8-3** Superimposed method for determining Coefficient of
Beam Utilization

## (5) **Estimate the Maintenance Factor**

Lighting efficiency is seriously impaired by lamp depreciation and by dirt on the reflecting and transmitting surfaces of the equipment. To compensate for the gradual depreciation of illumination on the floodlit area, a maintenance factor (MF) must be applied in the calculations to make allowance for the following:

a)  Loss of light output due to dirt on the lamp, reflector, and cover glass. Under comparable conditions, enclosed floodlights have a higher maintained efficiency than that of open units because the cover glass protects both the reflector and the lamp from the accumulation of dirt.

b)  Loss in light output of the lamp with life. Bulb blackening lowers floodlight efficiency, the reduction in beam lumens being about double the reduction in bare lamp output.

Maintenance factors are usually estimated to be from 0.65 to 0.85. However, if the floodlights are cleaned infrequently, or where lamps are replaced only on burnout, it is advisable to use lower maintenance factors. Difference in lumen maintenance among lamp types and sizes should also be considered. With narrow-beam floodlights, dirt on the reflector and cover glass tends to widen the beam spread, reducing the maximum candlepower more than the total light output. Thus for a small lighted area utilizing only the central part of a beam, a smaller percentage of the beam lumens will strike the target after the floodlight becomes dirty. Therefore, the depreciation in footcandle intensity will be greater than the depreciation in total light output, and it will be necessary to consider this in selecting a maintenance factor.

## (6) Determine the Number of Floodlights Required

$$\text{number of floodlights} = \frac{\text{area x footcandles}}{\text{beam lumens x CBU x MF}}$$

where area is the surface to be lighted in square feet; footcandles are as selected from Table 8-3. For the beam lumens, refer to manufacturers' catalogs for equipment under consideration. Where the supply voltage deviates from rated voltage, the lamp lumen output should be adjusted accordingly.

## (7) Determine the Location of the Floodlights

Physical limitations imposed by either the relationship of the building to its surrounding or by local regulations may drastically limit the design in the number of solutions or effects available. In general, four locations may be considered:

a)   on the building itself
b)   on adjacent ground
c)   on poles or ornamental standards
d)   on adjacent buildings

Floodlights located close to the surface to be lighted and aimed at a grazing angle will tend to emphasize the texture of the surface. Mounting the floodlights farther and farther from the surface de-emphasizes or flattens the texture and usually improves brightness uniformity.

The lamp and reflector should be shielded or louvered so that brightness can not be seen from any normal viewing locations. Floodlights should be located or shielded so that units do not light adjacent units thereby revealing their presence.

(8) **Check for Coverage and Uniformity**
After a tentative layout has been made, the uniformity may be checked by calculating the intensity of illumination at a few points on the floodlit surface. This may be done by the point-by-point method described in Chapter 2, using either a candlepower distribution curve or an isocandela diagram. If the uniformity is found to be unsatisfactory, a larger number of units may be required.

## 8.6 Application Guide

(1) **Buildings**
The floodlighting of a building is primarily a problem in esthetics, and therefore, each installation must be studied individually. Under some circumstances, particularly with small, utilitarian buildings or larger buildings that have no special architectural features, uniform illumination is desirable. To create the appearance of uniform brightness over the entire

facade of a building, it is necessary to increase the actual brightness appreciably toward the top.

With buildings of classical design or special architectural character, uniform illumination often defeats the purpose of the floodlighting, which should aim to preserve and bring out the architectural form. Buildings are primarily designed for daytime appearance, when the light comes from above. This effect is almost impossible to duplicate with floodlights, which must be mounted in nearby locations and usually at a height no greater than the elevation of the building. The only possible design is to achieve a result that is interesting and pleasing, yet different from the daytime appearance.

Shadows are essential to relief, and contrasts in brightness levels or sometimes in color can be used to bring out important details and to suppress others. Sculpture or architectural details needs particularly careful treatment to avoid flatness or grotesque shadows that may distort the appearance conceived by the architect.

(2) **Excavation and Construction**
Approximately 35-40,000 total lamp lumens in one to three floodlights will be required for each 5000 square feet of excavation area or for each 1000 to 2000 square feet of construction area. It is usually satisfactory to mount floodlights in groups of two or more on wood poles or towers 40 to 70 ft above ground. A minimum of two poles should be used, with enough poles on larger jobs to provide coverage at any working point from two or three floodlight banks. Fixed-pole spacing may be from 1-1/2 to 3 times mounting height, and as much as 5 times mounting height on large projects. Where a mechani-

cal shovel or crane is used, it is advisable to mount an automatic leveling floodlight on the boom.

(3) **Color**

Color can be provided in floodlighting installations in any one of several ways. Amber, blue, and red cover glasses are generally available for standard enclosed floodlights to replace the regular lens, or the floodlight may be recessed in a niche, the opening of which is covered with a filter. Where smaller amounts of color light are needed, 100 W PAR38 colortone (red, pink, yellow, green, blue, or blue-white) lamps are available, or a 300- or 500 W hard glass R-40 lamp may be used with a colored lens. Any color filter absorbs a large amount of light, and this loss must be considered in designing the installation.

## 8.7 Examples of Floodlighting Installation

The following examples of floodlighting installations which were designed by the author are illustrated here to highlight different themes and methods to achieve somewhat different aims in each case.

(1) **A Major Corporation Building**

In the past, incandescent lamps were the most useful and versatile source. However, from an energy efficiency and cost-saving standpoint, aesthetic lighting can be achieved with more efficient HID sources wherever feasible. A good example is the floodlighting of a corporation headquarters building. The building had been face-lifted with a marble chip finish. The design specified 10 luminaires fitted with 400 W HPS lamps mounted on the roof of a garage some 50 ft away from the building. Each unit was adjusted individually to a proper aiming angle to achieve a uniform illumination of 10 fc over the building face.

These floodlights have created a color of warmth and unmistakable identity, viewed by thousands of motorists traveling the Garden State Parkway at night. Figure 8-4 shows the building face being flood-lit with HPS luminaires.

Figure **8-4**  A corporate headquarter office building floodlighted with HPS light source

(2) **Pan Am Building**

The main faces (north and south) of this New York office building are floodlighted from the tenth floor setback up to the top of the 59th floor, a distance of 550 ft. To achieve such lighting, special searchlight-type luminaires were developed to house a specially designed incandescent lamp.

The lamp had to have the most compact filament possible to enable the luminaire mirrors to project narrow but intense beams to the top of the building. At the same time, enough spread is needed to enable the luminaires to cover the lower areas, and a burning life of more than a year is highly desirable. The lamp developed has a globular bulb of hard glass 8 in. in diameter and a special 80 V filament operating at 2000 W. A collector grid traps tungsten particles as the filament vaporizes, preventing bulb blackening and thereby assuring good lumen maintenance. The special luminaires have cast aluminum housing with mirrored glass reflectors and clear tempered lenses. Each of the 170 units produces 2,750,000 candelas.

The smaller faces of the building are floodlit with 1000 W quartz-iodine lamps in luminaires that supply the same light distribution provided by the searchlights on the larger sides. As shown in Figure 8-5 the floodlit building stands out with night city background.

(3) **A Lamp Manufacturing Plant**

The objectives are to light the front and ends of an office building brightly and flatly so that it would stand out against the background of the manufacturing building behind it, which, although less brightly lighted, is to have enough illumination to define its mass clearly. The office building is illuminated to a minimum intensity of 15 fc by lighting upward from ground level with weather proof fluorescent floodlights.

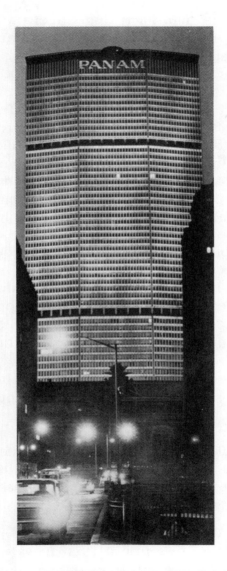

Figure **8-5** Pan Am building floodlighted with specially designed
luminaires and lamps

Special fluorescent lamps that have a low silhouette are used in the fixture.

For the manufacturing building, twenty seven 1500 W quartz-iodine floodlights are located at ground level, 14 ft from the wall on approximately 40 ft centers, to provide a pattern on the building face and create a "glow" as background for the highlighted office building and the shrubbery. To delineate the pattern of the facade, all floodlights are aimed at 45 degrees from perpendicular to the building face and upward to provide the effect of the sun shining on the building at a 45-degree angle. One floodlight is opposite each column and each corner, with the others located as required to give adequate overlapping of patterns. This installation was one of the earliest application of the newly developed 1500 W quartz-iodine floodlight luminaires.

The shrubbery surrounding the office building is illuminated by batteries of floodlights on the building roof along both ends. The best color for the shrubs can be provided by mercury lighting, so 100 W PAR38 mercury lamps are chosen. They are mounted in heavy-duty cast-aluminum outdoor luminaires, which are neat and small for good appearance, and so aimed that they do not produce glare toward viewers of the scene. Figure 8-6 shows a brightly floodlit office building in the foreground of the manufacturing building. Figure 8-7 shows a fluorescent lamp manufacturing plant building that has sunset pattern floodlighting.

(4) **A Water Tower**

Floodlighting a sign on a circular water tower, without spill or scatter, requires rather more than rule-of-thumb design. The light sources chosen had to be an economical lamp which would produce a narrow

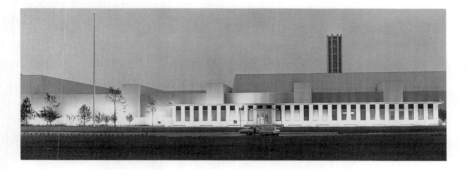

Figure **8-6**  A brightly floodlit office building

Figure **8-7**  A lamp manufacturing plant building floodlit with
a sunset pattern design

searchlight type of beam and project all of the light
exactly where it was aimed.

In this case, low-voltage 120 W PAR64 very-
narrow-spot lamps accomplished the task. The bottom part of the tank and the upper tower legs were
silhouetted with 500 W PAR64 narrow-spot lamps
operating on a 120 V circuit. The 120 W PAR64
lamps operate on a 6 V circuit and can produce a
controlled pencil-thin beam which is very effective for
uplighting buildings, columns, steeples, towers, and
the like. Its beam spread is 4-1/2 degrees by

7 degrees and is achieved by masking critical areas of the reflector to prevent refocusing of the light. A matte black light shield produces an extremely sharp beam cutoff. Mounted precisely in relation to the reflector's focal point, a specially designed filament yields a maximum beam candlepower of 170,000. Figure 8-8 shows the candlepower distribution of the 120 W PAR64 very-narrow-spot lamp and the 500 W PAR64 narrow-spot lamp.

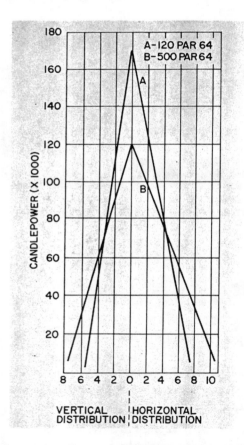

Figure **8-8**  Candlepower distribution curves of 120 W and
500 W PAR64 lamps

The water tank stands 100 ft above the grade. The overall diameter of the tank is 22 ft. The corporate sign is approximately 12 ft in diameter and 120 ft above grade. The lighting banks pepper the sign from two sides: from the nearby building roof and from an erected pole. Each is located approximately 30 ft above the ground. The bank located on a nearby building roof consists of six 120 W and three 500 W PAR64 lamps. The bank located on the top of a new pole consists of four 120 W and two 500 W PAR64 lamps. Figure 8-9 is a sketch of the scheme with all essential dimensions of the tower and mounting locations from both field measurements and calculations.

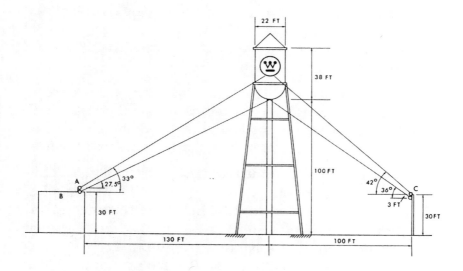

Figure **8-9** Dimensions and locations of floodlights for a water tower

All lamps are contained in a weatherproof cast-aluminum unit equipped with clear high-heat-treated tempered lenses, and adjustable cast arms permit exact aiming of the units. Raintight aluminum wireways are used to connect the floodlights at each bank, where the lower row of luminaires is aimed at the bottom of the tank and the upper row at the sign. Figure 8-10 shows the water tower floodlighted by PAR64 lamps as the light source. The wiring scheme for the entire installation is shown in Figure 8-11. A three-pole starter is used to control power supply for the floodlights, which can be turned on and off by an astronomical clock. The main disconnect switch, line starter, timer, and step-down transformer are installed in a nearby shed so that indoor equipment can be used. 120 V feeders are enclosed in steel conduit, while 12 V circuits in the form of underground cables are directly buried in the ground from the control station to the pole.

The following additional examples are cited here to emphasize a recent trend, which is to use the highly efficient HID light source instead of the traditionally popular source to achieve esthetics as well as energy/cost savings:

(1) **Indiana State Capital Building** – Originally designed for 35 fc, the dome's existing 56-1500 W incandescent filament floodlights have been successfully replaced with 56-400 W HPS floodlights. The new system reduced its lighting load from 85 kW to 27 kW. With an average of 3000 burning hours a year and at a cost of $0.03/kWh, the annual energy/cost savings would amount to $5,200.

(2) **The Haywood City Center** – a modern high-rise structure. The architect considered the nighttime appearance of the building in his design, but the budget

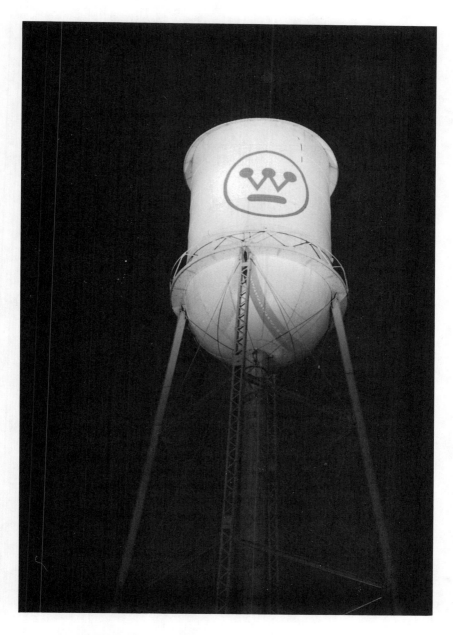

Figure **8-10** A water tower floodlit with PAR64 light source

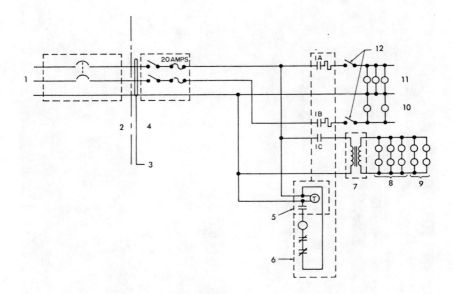

1— 230/115-volt, three-wire, 60-cps service.
2— Existing service line in shed.
3— 3/4-inch—3 No. 10 RHW.
4— Safety switch mounted in shed.
5— Tork timer to control on and off.
6— Line starter.
7— 120/12-volt transformer, 1 1/2-kVa, mounted in shed.
8— Six 120-watt, six-volt PAR64 very narrow spots mounted on roof.
9— Four 120-watt, six-volt PAR64 narrow spots on new pole.
10— Two 500-watt, 120-volt PAR64 narrow spots on new pole.
11— Three 500-watt, 120-volt PAR64 narrow spots on roof.
12— S.P. disconnect switch, outdoors at pole and on nearby shed.

Figure **8-11** Wiring scheme for floodlighting a water tower constraints placed a severe damper on the possibility of floodlighting. The solution to this problem was to provide a combined floodlighting system, which consists of as many 1000 W metal halide luminaires as the budget could allow and low initial cost 1500 W tungsten-halogen luminaires. The metal halide luminaires would provide light on the middle and lower portion of the structure. The punch from narrow-beam tungsten-halogen units was used to reach the top of the tall structure. The color shift from cooler metal halide on the lower half to the warmer tungsten-halogen lamps on the upper half turned out to be not objectionable.

The combination system produces approximately 12 fc of vertical illumination on the building. This exceeds the IES recommended level of 10 fc for a building with 45 to 70% reflectance in a dark surrounding. The result was a uniform floodlighting in intensity and color.

## Bibliography

Chen, Kao, Floodlighting Technique for A Water Tower, Illuminating Engineering, May 1967, pp. 305-307.

Chen, Kao, and Karns, E.B., New Techniques Enhance Effectiveness of Building Floodlighting, Westinghouse Engineer, July 1968, pp. 118-121.

Chen, Kao, What's New in Floodlighting, IEEE Transactions on Industry Applications, July/Aug. 1977, pp. 343-347.

Floodlighting with High-Pressure-Sodium, Data Sheet, Lighting Design and Application, June 1972.

Lighting Handbook, N.A. Philips Lighting Company, Somerset, New Jersey, 1983.

# Chapter 9

# Lighting and Energy Standards

## 9.1 Chronological Development of the Energy Standards

Early in 1974, the document "Energy Conservation Guidelines for Existing Office Buildings" was published by the General Services Administration and the Public Building Services (GSA/PBS). At approximately the same time, the National Bureau of Standards (NBS) prepared the document "Design and Evaluation Criteria for Energy Conservation in New Buildings" for the National Conference of States on Building Codes and Standards (NCSBCS) covering lighting for new construction. NCSBCS asked the American Society of Heating, Refrigeration, and Air Conditioning Engineers (ASHRAE) to prepare a standard on energy conservation for new buildings based on this NBS document. ASHRAE turned to the Illuminating Engineering Society (IES), and the IES Task Committee on Energy Budgeting Procedures was formed. The work of this committee culminated in recommendations that were published by ASHRAE as Chapter 9 of the ASHRAE 90-75 and by IES as "IES Recommended Lighting Power Budget Determination Procedures EMS-1". NCSBCS also adopted it as part of

169

its Model Building Code. Both documents established a procedure for determining a "lighting power budget".

In 1975, Congress passed Public Law 94-163, entitled "Energy Policy and Conservation Act of 1975", which was amended by Public Law 94-385, "Energy Conservation Standards for New Building Act of 1976". Public Law 94-164 makes mandatory certain lighting efficiency standards set forth in Public Laws 94-163 and 94-385.

In 1976, the Energy Research and Development Association (ERDA) contracted with NCSBCS to codify ASHRAE 90-75. The resulting document was called "The Model Code for Energy Conservation in New Building", or more simply, the "Model Code". The Model Code has been adopted by a number of states to satisfy the requirements of Public Laws 94-163 and 94-385. In June 1976, the IES Board of Directors adopted important revisions to Chapter 9 of ASHRAE 90-75 that tightened the lighting power budget procedure to assure energy conservation. These revisions were included in 90-75R but did not become part of the NCSBCS Model Code.

ASHRAE 90-75 was cosponsored by IES and ASHRAE and submitted to the American National Standards Institute (ANSI) in late 1977 for adoption as an ANSI standard. The document that resulted was known as ANSI/ASHRAE/IES Standard 90, "Energy Conservation in New Buildings".

There have been several revisions on the ANSI/ASHRAE/IES 90-75R since then. All were included in the lighting portion of ANSI/ASHRAE/IES 90A-1980, "Energy Conservation in New Building Design", and in EMS-1-1981, "IES Recommended Lighting Power Budget Determination Procedure".

ANSI/ASHRAE/IES Standard 90-75R was by far the most popular energy conservation design standard. It has been adopted as energy codes by most states within the United States and serves as a model standard in many other countries. Although this standard underwent a revision in 1980, major revisions were not made to incorporate new practices and technology.

ASHRAE/IES 90.1-1989, "Energy Efficient Design of New Buildings except New Low-Rise Residential Buildings" is the third generation document on building energy efficiency since the first publication in 1975. It sets forth design requirements for the efficient use of energy in new buildings intended for human occupancy. The requirements apply to the building envelope, distribution of energy, system and equipment for auxiliaries, heating, ventilating, air-conditioning, service water heating, lighting, and energy management. This standard is intended to be a voluntary standard which can be adopted by building officials for state and local codes. Three parallel and alternative paths for compliance are provided: prescriptive, system performance, and building energy cost budget methods. Each of these methods may be used selectively and interchangeably when determining building subsystems compliance. The interaction between the three compliance methods fulfills needs that arise during the various phases of the building process.

## 9.2 Development of the Lighting Energy Standards

In 1975, work was begun by IES on a series of six documents that deal with energy standards for existing buildings. The documents cover low-rise residential, high-rise residential, commercial, industrial, institutional, and public assembly occupancies. Several of these documents have since become ANSI standards.

As a means for simplifying and shortening the EMS-1 procedure, the IES developed a Unit Power Density (UPD) procedure and published it as EMS-6 in 1980. The present LEM-1-1982, "Lighting Power Limit Determination", is a further refinement which combines EMS-6 and EMS-1 and, as such, supersedes both. LEM-1 is concerned only with the determination of lighting power limit; lighting control guidelines are contained in LEM-3, "Design Considerations for Effective Building Lighting Energy Utilization".

Since 1982, IES has published the LEM series. In addition to LEM-1 and LEM-3 are LEM-2, "Lighting Energy Limit Determination", LEM-4, "Energy Analysis of Building Lighting Design and Installation", and LEM-6, "IES Guidelines for Unit Power Density (UPD) for New Roadway Lighting Installation".

## 9.3 Energy Policy Act

On October 25, 1992, the Energy Policy Act was signed into law by the President. Among the many provisions, this act establishes energy efficiency standards for HVAC, lighting, and motor equipment; encourages establishment of a national window energy-efficiency rating system; and encourages state regulators to pursue demand-side-management (DSM) programs.

Under this bill, lighting manufacturers will have 3 years to stop making F96T12 and F96T12/HO 8-ft fluorescent lamps and some types of incandescent reflectors. Standard F40 lamps except in the SP and SPX or equivalent types of high color rendering lamps, would also fade away. General service incandescent lamps to be axed would include those from 30 to 100 W, in 115 to 130 V ratings, having medium screw bases, of both reflector and PAR types, having a diameter larger than 2-3/4 inches.

There are no immediate regulations impacting HID lamps. Within 18 months of the legislation's enactment, the DOE will determine the HID types for which standards could possibly save energy and publish testing requirements for these lamps.

As far as the general service lamps are concerned, the most common incandescent lamps – 40 W, 60 W, 75 W, 100 W, and 150 W – are not covered by an efficiency standard because there is no suitable method to ensure energy savings. These types, however, are covered by another provision of the law, namely the energy efficiency labeling standards.

Effective April 28, 1994, the Federal Trade Commission (FTC) must provide manufacturers with labeling requirements for all lamps covered: fluorescent, incandescent, and reflector incandescent. Though not yet defined, the proposals include: an energy rating for the lamps, probably LPW (lumens per watt), and energy cost per year to operate the lamp. For standardization, the cost per year is likely to be based on an operating cost of 10 cents per kWh with four hrs/day of operation. The energy efficiency label will then allow side-by-side comparison of two different lamp types, thus enabling consumers to make a more intelligent choice of lamps; taking into account not just the purchase price, but also the operating cost. Manufacturers must begin applying labels by April 28, 1995.

Table 9-1 shows the proposed efficiency standards for the fluorescent lamps, and Table 9-2 shows the proposed efficiency standards for incandescent reflector lamps.

There is no requirement to replace all existing lamps in any installation. However, as these lamps burn out, the replacement must meet the new standards.

Replacement for popular fluorescent types includes reduced-wattage energy saving types. These lamps will meet the color and efficiency standards, so will the full wattage trisphosphor lamps having a CRI over 69. On the incandescent side, replacements for standard incandescent spot and flood lamps will be lower wattage halogen type reflector lamps which do meet the LPW requirements. The halogen and halogen/infrared types of reflector lamps will remain the only type of such lamps on the market. ER and BR types, those intended for rough or vibration service will also be excluded here.

In carrying out the replacement standards, expected savings on the lighting portion of this legislation will amount to 4,300 MW in power and 37 billion kWh in energy usage. These huge savings are equal to the annual electric energy use of more than 4 million homes.

The earliest provision to the new law becomes effective at the end of April, 1994. Effective that date, manufacturers can no longer produce 8-ft lamps which do not comply with the efficiency standards. Effective October 31, 1995, lamps covered include 4-ft fluorescents, 2-ft U-shaped fluorescents, and incandescent reflector types.

There is also a provision for lighting-fixture manufacturers to come up with voluntary luminaire efficiency standards. If these standards are found to be inadequate, the Department of Energy will come up with the mandatory efficiency standards.

The new Energy Policy Act is all encompassing. It promises to change forever the way the industries produce, distribute, and utilize the valued energy resources. The end result should be increased energy security, decreased environmental emissions, and cleaner air and water for all mankind.

Table **9-1** The proposed efficiency standards for fluorescent lamps

| Lamp type | Nominal lamp wattage | Minimum average CRI | Minimum average lamp efficacy |
|---|---|---|---|
| F40 | >35W | 69 | 75 |
| F40 | </=35W | 45 | 75 |
| F40/U | >35W | 69 | 68 |
| F40/U | </=35W | 45 | 64 |
| F96T12 | >65W | 69 | 80 |
| F96T12 | </=65W | 45 | 80 |
| F96T12/HO | >100W | 69 | 80 |
| F96T12/HO | </=100W | 45 | 80 |

The above excludes lamps designed for plant growth, cold temperature service, reflectorized/aperture, impact resistance, reprographic service, colored lighting, ultraviolet and lamps with CRI of more than 82.

Table **9-2** The proposed efficiency standards for incandescent reflector lamps

| Nominal lamp wattage | Minimum average lamp efficacy (LPM) |
|---|---|
| 40-50 | 10.5 |
| 51-66 | 11.0 |
| 67-85 | 12.5 |
| 86-115 | 14.0 |
| 116-155 | 14.5 |

The above excludes miniature, decorative, traffic signal, marine, mine, stage/studio, railway, colored lamps, and other special application types.

# Index